INTRODU
TO
BESSEL
FUNCTIONS

By Frank Bowman

Dover Publications Inc.
New York

This new Dover edition first published in 1958, is an unabridged and unaltered republication of the first edition.

Standard Book Number: 486-60462-4
Library of Congress Catalog Card Number: 58-11271

Manufactured in the United States of America

Dover Publications, Inc.
180 Varick Street
New York, N.Y. 10014

PREFACE

THIS small book has grown out of lectures given from time to time in the College of Technology, Manchester. It is hoped that it may serve as an introduction to the larger treatises on Bessel functions and their applications.

The author is indebted to Dr. W. N. Bailey for reading the manuscript, and Mr. H. Tilsley for drawing the figures, and particularly to Dr. S. Verblunsky for reading the manuscript and proofs and for many helpful suggestions.

F. B.

CONTENTS

TABLE I

x.	$J_0(x)$.	$J_1(x)$.
0	+1	0
1	+ ·7652	+ ·4401
2	+ ·2239	+ ·5767
2·405	0	+ ·5191
3	− ·2601	+ ·3391
3·832	− ·4028	0
4	− ·3971	− ·0660
5	− ·1776	− ·3276
5·520	0	− ·3403
6	+ ·1506	− ·2767
7	+ ·3001	− ·0047
7·016	+ ·3001	0
8	+ ·1717	+ ·2346
8·654	0	+ ·2715
9	− ·0903	+ ·2453
10	− ·2459	+ ·0435
10·173	− ·2497	0
11	− ·1712	− ·1768
11·792	0	− ·2325
12	+ ·0477	− ·2234
13	+ ·2069	− ·0703
13·324	+ ·2184	0
14	+ ·1711	+ ·1334
14·931	0	+ ·2065

TABLE II

ROOTS OF THE EQUATION $J_n(x) = 0$

	$J_0 = 0$.	$J_1 = 0$.	$J_2 = 0$.	$J_3 = 0$.	$J_4 = 0$.	$J_5 = 0$.
1	2·405	3·832	5·136	6·380	7·588	8·771
2	5·520	7·016	8·417	9·761	11·065	12·339
3	8·654	10·173	11·620	13·015	14·373	15·700
4	11·792	13·324	14·796	16·223	17·616	18·980
5	14·931	16·471	17·960	19·409	20·827	22·218

A few of the more modern books on Bessel Functions are given below ; a comprehensive bibliography will be found at the end of Watson's " Theory of Bessel Functions ".

Watson, " Theory of Bessel Functions " (Cambridge, 1922).

Gray and Mathews, " Bessel Functions " (2nd edn., London, 1922).

Nielsen, " Handbuch d. Theorie d. Cylinderfunktionen " (Leipzig, 1904).

Schafheitlin, " Die Theorie d. Besselschen Funktionen " (Leipzig, 1908).

McLachlan, " Bessel Functions for Engineers " (Oxford, 1934).

Weyrich, " Die Zylinderfunktionen und ihre Anwendungen " (Leipzig 1937).

CHAPTER I

BESSEL FUNCTIONS OF ZERO ORDER

§ 1. *Bessel's function of zero order.*

The function known as Bessel's function of zero order, and denoted by $J_0(x)$, may be defined by the infinite power-series

$$J_0(x) = 1 - \frac{x^2}{2^2} + \frac{x^4}{2^2 . 4^2} - \frac{x^6}{2^2 . 4^2 . 6^2} + \dots \qquad (1.1)$$

If u_r denotes the rth term of this series, we have

$$\frac{u_{r+1}}{u_r} = - \frac{x^2}{(2r)^2}$$

which $\to 0$ when $r \to \infty$, whatever the value of x. Consequently, the series converges for all values of x, and since it is a power-series, the function $J_0(x)$ and all its derivatives are continuous for all values of x, real or complex.

§ 2. *Bessel's function of order n, when n is a positive integer.*

The function $J_n(x)$, known as Bessel's function of order n, may be defined, when n is a positive integer, by the infinite power-series

$$J_n(x) = \frac{x^n}{2^n n!}\left(1 - \frac{x^2}{2 . 2n+2} + \frac{x^4}{2 . 4 . 2n+2 . 2n+4} - \dots\right) \qquad (1.2)$$

which converges for all values of x, real or complex.

In particular, when $n = 1$ we have

$$J_1(x) = \frac{x}{2} - \frac{x^3}{2^2 . 4} + \frac{x^5}{2^2 . 4^2 . 6} - \frac{x^7}{2^2 . 4^2 . 6^2 . 8} + \dots \qquad (1.3)$$

and when $n = 2$

$$J_2(x) = \frac{x^2}{2.4} - \frac{x^4}{2^2.4.6} + \frac{x^6}{2^2.4^2.6.8} - \frac{x^8}{2^2.4^2.6^2.8.10} + \dots \quad (1.4)$$

We note that $J_n(x)$ is an even function of x when n is even, odd when n is odd.

The graphs of $J_0(x)$, $J_1(x)$ are indicated in Fig. 1.

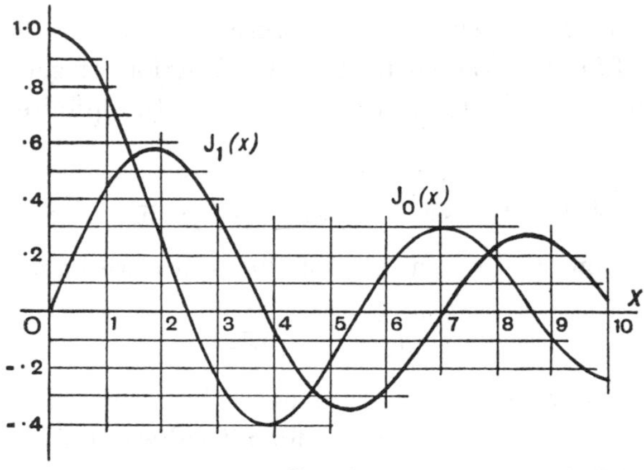

FIG. 1.

Extensive tables of values of $J_n(x)$, especially of $J_0(x)$ and $J_1(x)$, have been calculated on account of their applications to physical problems.*

§ 3. *Bessel's equation of zero order.*

By differentiating the series for $J_0(x)$ and comparing the result with the series for $J_1(x)$, we find †

$$\frac{dJ_0(x)}{dx} = -J_1(x). \quad . \quad . \quad . \quad (1.5)$$

* See Watson: "Theory of Bessel Functions"; Gray and Mathews: "Bessel Functions"; Jahnke und Emde: "Funktionentafeln"; Dale: "Five-Figure Tables."

† Cf. $\frac{d}{dx}(\cos x) = -\sin x$.

Again, after multiplying the series for $J_1(x)$ by x and differentiating, we find

$$\frac{d}{dx}\{xJ_1(x)\} = xJ_0(x). \qquad (1.6)$$

Using (1.5), we can write (1.6) in the form

$$\frac{d}{dx}\left(x\frac{dJ_0(x)}{dx}\right) + xJ_0(x) = 0, \qquad (1.7)$$

or

$$x\frac{d^2J_0(x)}{dx^2} + \frac{dJ_0(x)}{dx} + xJ_0(x) = 0. \qquad (1.8)$$

Thus $y = J_0(x)$ satisfies the linear differential equation of the second order

$$\frac{d}{dx}\left(x\frac{dy}{dx}\right) + xy = 0, \qquad (1.9)$$

or

$$x\frac{d^2y}{dx^2} + \frac{dy}{dx} + xy = 0, \qquad (1.10)$$

or

$$\frac{d^2y}{dx^2} + \frac{1}{x}\frac{dy}{dx} + y = 0, \qquad (1.11)$$

which is known as *Bessel's equation of zero order.*

§ 4. *Bessel functions of the second kind of zero order.*

A solution of Bessel's equation which is not a numerical multiple of $J_0(x)$ is called a *Bessel function of the second kind.* Let u be such a function, and let $v = J_0(x)$; then, by (1.10),

$$xu'' + u' + xu = 0,$$
$$xv'' + v' + xv = 0.$$

Multiplying the first of these equations by v and the second by u and subtracting, we have

$$x(u''v - uv'') + u'v - uv' = 0,$$

which, since $\qquad u''v - uv'' \equiv \frac{d}{dx}(u'v - uv'),$

can be written $\qquad \frac{d}{dx}\{x(u'v - uv')\} = 0.$

Hence $$x(u'v - uv') = B,$$
where B is a constant. Dividing by xv^2, we have
$$\frac{u'v - uv'}{v^2} = \frac{B}{xv^2}$$
that is,
$$\frac{d}{dx}\left(\frac{u}{v}\right) = \frac{B}{xv^2}$$
and hence, by integration,
$$\frac{u}{v} = A + B\int\frac{dx}{xv^2}.$$

Consequently, since $v = J_0(x)$,
$$u = AJ_0(x) + BJ_0(x)\int\frac{dx}{xJ_0{}^2(x)} \quad . \quad (1.12)$$
where A, B are constants, and $B \neq 0$ since u is not a constant multiple of $J_0(x)$, by definition.

§ 5. If, in the last integral, $J_0(x)$ is replaced by its series, and the integrand expanded in ascending powers of x, we find for the first few terms
$$\frac{1}{xJ_0{}^2(x)} = \frac{1}{x} + \frac{x}{2} + \frac{5x^3}{32} + \ldots$$
and therefore
$$J_0(x)\int\frac{dx}{xJ_0{}^2(x)} = J_0(x)\left(\log x + \frac{x^2}{4} + \frac{5x^4}{128} + \ldots\right)$$
$$= J_0(x)\log x + \left(1 - \frac{x^2}{2^2} + \ldots\right)\left(\frac{x^2}{4} + \frac{5x^4}{128} + \ldots\right)$$
$$= J_0(x)\log x + \frac{x^2}{4} - \frac{3x^4}{128} + \ldots$$

Consequently, if we put
$$Y_0(x) = J_0(x)\log x + \frac{x^2}{4} - \frac{3x^4}{128} + \ldots \quad (1.13)$$
then $Y_0(x)$ is a particular Bessel function of the second kind; it is called *Neumann's Bessel function of the second kind of zero order;* the general term in its expansion can be obtained by other methods (§ 8).

Since $J_0(x) \to 1$ when $x \to 0$, it follows from (1.13) that $Y_0(x)$ behaves like $\log x$ when x is small, and hence that $Y_0(x) \to -\infty$ when $x \to +0$.

§ 6. It follows from (1.12) that every Bessel function of the second kind of zero order can be written in the form

$$AJ_0(x) + BY_0(x).$$

The one that has been most extensively tabulated is Weber's,* which is denoted by $\mathrm{Y}_0(x)$ and is obtained by putting

$$A = -\frac{2}{\pi}(\log 2 - \gamma), \quad B = \frac{2}{\pi}$$

and hence

$$\mathrm{Y}_0(x) = \frac{2}{\pi}\{Y_0(x) - (\log 2 - \gamma)J_0(x)\}, \quad . \quad (1.14)$$

where γ denotes Euler's constant, defined by

$$\gamma = \lim_{n\to\infty}\left(1 + \frac{1}{2} + \frac{1}{3} + \ldots + \frac{1}{n} - \log n\right) = 0{\cdot}5772\ldots \quad (1.15)$$

We note that, when x is small,

$$\mathrm{Y}_0(x) = \frac{2}{\pi}\{\log x - (\log 2 - \gamma)\ \ldots\} \quad . \quad (1.16)$$

the remaining terms being small in comparison with unity.

As far as applications are concerned, it is usually sufficient to bear in mind that $\mathrm{Y}_0(x)$ is a Bessel function of the second kind whose values have been tabulated; that x must be positive for $\mathrm{Y}_0(x)$ to be real, on account of the term involving $\log x$ in (1.13); and that $\mathrm{Y}_0(x) \to -\infty$ when $x \to +0$.

The graphs of $J_0(x)$ and $\mathrm{Y}_0(x)$ are shown together in Fig. 2.

§ 7. *General solution of Bessel's equation of zero order.*

Since $J_0(x)$ and $\mathrm{Y}_0(x)$ are independent solutions of the equation

$$\frac{d^2y}{dx^2} + \frac{1}{x}\frac{dy}{dx} + y = 0,$$

* Watson, § 3.54.

the general solution can be written

$$y = AJ_0(x) + BY_0(x), \quad . \quad . \quad (1.17)$$

where A, B are arbitrary constants, and $x > 0$ for $Y_0(x)$ to be real.

If we replace x by kx, where k is a constant, the equation becomes

$$\frac{1}{k^2}\frac{d^2y}{dx^2} + \frac{1}{kx}\frac{dy}{kdx} + y = 0.$$

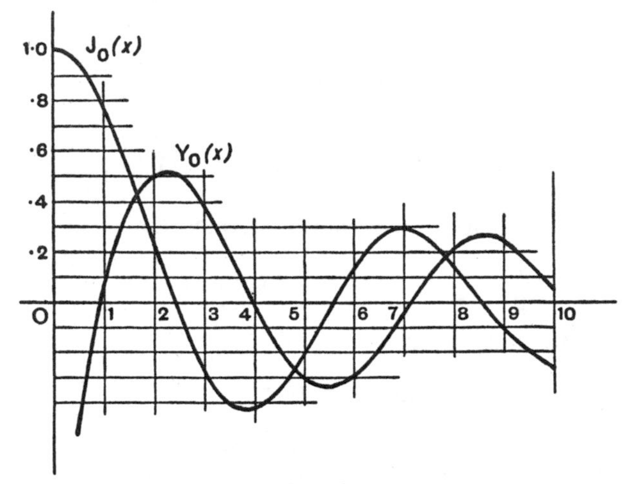

Fig. 2.

Multiplying by k^2, we deduce that the general solution of the equation

$$\frac{d^2y}{dx^2} + \frac{1}{x}\frac{dy}{dx} + k^2y = 0 \quad . \quad . \quad (1.18)$$

can be written

$$y = AJ_0(kx) + BY_0(kx) \quad . \quad . \quad (1.19)$$

where $k > 0$ for $Y_0(kx)$ to be real when $x > 0$.

§ 8. *The general solution by Frobenius's method.*

Bessel's equation belongs to the type to which Frobenius's method of solution in series can be applied. Put

$$Ly \equiv \left(x\frac{d^2}{dx^2} + \frac{d}{dx} + x\right)y \quad . \quad . \quad . \quad (1.20)$$

and make the substitution

$$y = x^\rho + c_1 x^{\rho+1} + c_2 x^{\rho+2} + c_3 x^{\rho+3} + \ldots \qquad (1.21)$$

We obtain, after collecting like terms,

$$\begin{aligned} Ly = \rho^2 x^{\rho-1} + (\rho+1)^2 c_1 x^\rho + \{(\rho+2)^2 c_2 + 1\} x^{\rho+1} \\ + \{(\rho+3)^2 c_3 + c_1\} x^{\rho+2} + \ldots \qquad (1.22) \end{aligned}$$

Now let $c_1, c_2, c_3, \ldots$ be chosen to satisfy the equations

$$\begin{aligned} &(\rho+1)^2 c_1 = 0, \\ &(\rho+2)^2 c_2 + 1 = 0, \\ &(\rho+3)^2 c_3 + c_1 = 0, \ldots \end{aligned}$$

Then, unless ρ is a negative integer,

$$\begin{aligned} c_1 &= c_3 = c_5 = c_7 = \ldots = 0, \\ c_2 &= -\frac{1}{(\rho+2)^2}, \\ c_4 &= -\frac{c_2}{(\rho+4)^2} = \frac{1}{(\rho+2)^2(\rho+4)^2} \ldots \end{aligned}$$

Substituting these values in (1.21) and (1.22), we deduce that, if

$$y = x^\rho \left\{1 - \frac{x^2}{(\rho+2)^2} + \frac{x^4}{(\rho+2)^2(\rho+4)^2} - \ldots\right\} \qquad (1.23)$$

and if ρ is not a negative integer, then

$$x\frac{d^2y}{dx^2} + \frac{dy}{dx} + xy = \rho^2 x^{\rho-1}. \qquad (1.24)$$

Putting $\rho = 0$ in (1.23) and (1.24) we see again that

$$y = J_0(x) = 1 - \frac{x^2}{2^2} + \frac{x^4}{2^2 . 4^2} - \frac{x^6}{2^2 . 4^2 . 6^2} + \ldots$$

is a solution of Bessel's equation

$$x\frac{d^2y}{dx^2} + \frac{dy}{dx} + xy = 0.$$

Further, differentiating (1.24) partially with respect to ρ, we get

$$x\frac{d^2}{dx^2}\left(\frac{\partial y}{\partial \rho}\right) + \frac{d}{dx}\left(\frac{\partial y}{\partial \rho}\right) + x\left(\frac{\partial y}{\partial \rho}\right) = 2\rho x^{\rho-1} + \rho^2 x^{\rho-1} \log x,$$

and hence, when $\rho = 0$,

$$x\frac{d^2}{dx^2}\left(\frac{\partial y}{\partial \rho}\right)_0 + \frac{d}{dx}\left(\frac{\partial y}{\partial \rho}\right)_0 + x\left(\frac{\partial y}{\partial \rho}\right)_0 = 0,$$

from which it follows that $(\partial y/\partial \rho)_{\rho=0}$ is a second solution. Now, from (1.23),

$$\frac{\partial y}{\partial \rho} = x^\rho \log x \left\{1 - \frac{x^2}{(\rho+2)^2} + \frac{x^4}{(\rho+2)^2(\rho+4)^2} - \ldots\right\}$$
$$+ x^\rho \left\{\frac{2x^2}{(\rho+2)^2}\frac{1}{\rho+2} - \frac{2x^4}{(\rho+2)^2(\rho+4)^2}\left(\frac{1}{\rho+2} + \frac{1}{\rho+4}\right)\right.$$
$$\left. + \frac{2x^6}{(\rho+2)^2(\rho+4)^2(\rho+6)^2}\left(\frac{1}{\rho+2} + \frac{1}{\rho+4} + \frac{1}{\rho+6}\right) - \ldots\right\}.$$

Hence, putting $\rho = 0$ and $Y_0(x) = (\partial y/\partial \rho)_{\rho=0}$ we obtain the second solution

$$Y_0(x) = J_0(x)\log x + \frac{x^2}{2^2} - \frac{x^4}{2^2 \,.\, 4^2}(1 + \tfrac{1}{2}) + \frac{x^6}{2^2 \,.\, 4^2 \,.\, 6^2}(1 + \tfrac{1}{2} + \tfrac{1}{3}) - \ldots \quad (1.25)$$

which is Neumann's Bessel function of the second kind of zero order, in a form which indicates the general term (§ 5).

It follows that the general solution of the equation can be written

$$y = AJ_0(x) + BY_0(x),$$

which is equivalent to (1.17).

§ 9. To examine the convergence of the series that follows $J_0(x) \log x$ in (1.25), we can put, by (1.15),

$$1 + \frac{1}{2} + \frac{1}{3} + \ldots + \frac{1}{n} = \log n + \gamma + \epsilon_n, \quad . \quad (1.26)$$

where $\epsilon_n \to 0$ when $n \to \infty$. Hence if u_r denote the rth term of the series, we have

$$\frac{u_r}{u_{r-1}} = -\frac{x^2}{(2r)^2}\frac{\log r + \gamma + \epsilon_r}{\log(r-1) + \gamma + \epsilon_{r-1}},$$

which $\to 0$ when $r \to \infty$, whatever the value of x. Consequently, the series converges absolutely for all values of x, real or complex.

§ 10. *Integrals.*

We notice next certain integrals involving Bessel functions in their integrands. Firstly, from (1.5) and (1.6) we have

$$\int J_1(x)dx = -J_0(x), \quad . \quad . \quad (1.27)$$
$$\int xJ_0(x)dx = xJ_1(x). \quad . \quad . \quad (1.28)$$

Secondly, we note that the indefinite integral

$$\int J_0(x)dx \quad . \quad . \quad . \quad . \quad (1.29)$$

cannot be expressed in a simpler form, but on account of its importance the value of the definite integral

$$\int_0^x J_0(t)dt \quad . \quad . \quad . \quad (1.30)$$

has been tabulated.*

Thirdly, we shall obtain a reduction formula for the integral

$$\int x^n J_0(x)dx. \quad . \quad . \quad . \quad (1.31)$$

Put $\qquad u_n = \int x^n J_0(x)dx = \int x^{n-1}d\{xJ_1(x)\},$

by (1.6). Then, integrating by parts, we have

$$\begin{aligned} u_n &= x^{n-1} \,.\, xJ_1(x) - \int xJ_1(x) \,.\, (n-1)x^{n-2}dx \\ &= x^n J_1(x) + (n-1)\int x^{n-1}dJ_0(x), \end{aligned}$$

by (1.5); and on integrating by parts again,

$$u_n = x^n J_1(x) + (n-1)x^{n-1}J_0(x) - (n-1)^2\int x^{n-2}J_0(x)dx,$$

that is,

$$u_n = x^n J_1(x) + (n-1)x^{n-1}J_0(x) - (n-1)^2 u_{n-2} \quad (1.32)$$

which is the reduction formula required.

It follows that, if n is a positive integer, the integral (1.31) can be made to depend upon (1.28) if n is odd, or upon (1.29) if n is even.

§ 11. *Lommel's integrals.*

Put $u = J_0(\alpha x)$, $v = J_0(\beta x)$, where α, β are constants; then, by § 7, writing $u' = du/dx$, $u'' = d^2u/dx^2$, . . . we have

$$xu'' + u' + \alpha^2 xu = 0, \quad . \quad . \quad (1.33)$$

$$xv'' + v' + \beta^2 xv = 0. \quad . \quad . \quad (1.34)$$

Multiplying the first of these equations by v and the second by u and subtracting, we get

$$x(u''v - uv'') + (u'v - uv') = (\beta^2 - \alpha^2)xuv,$$

that is, $\qquad \dfrac{d}{dx}\{x(u'v - uv')\} = (\beta^2 - \alpha^2)xuv,$

and hence, by integration,

$$(\beta^2 - \alpha^2)\int xuvdx = x(u'v - uv'),$$

* Watson, p. 752.

and therefore

$$(\beta^2 - \alpha^2)\int xJ_0(\alpha x)J_0(\beta x)dx$$
$$= x\{\alpha J_0'(\alpha x)J_0(\beta x) - \beta J_0'(\beta x)J_0(\alpha x)\}, \quad (1.35)$$

since $u' = \alpha J_0'(\alpha x)$, $v' = \beta J_0'(\beta x)$.

Again, multiplying (1.33) throughout by $2xu'$, we have

$$2xu'\frac{d}{dx}(xu') + 2\alpha^2x^2uu' = 0,$$

or

$$\frac{d}{dx}(x^2u'^2 + \alpha^2x^2u^2) - 2\alpha^2xu^2 = 0.$$

Integrating, we get

$$x^2u'^2 + \alpha^2x^2u^2 = 2\alpha^2\int xu^2dx,$$

and hence

$$2\alpha^2\int xJ_0^2(\alpha x)dx = \alpha^2x^2J_0^2(\alpha x) + x^2\alpha^2J_0'^2(\alpha x),$$

and therefore

$$\int xJ_0^2(\alpha x)dx = \tfrac{1}{2}x^2\{J_0^2(\alpha x) + J_1^2(\alpha x)\}. \quad . \quad (1.36)$$

In particular, when we integrate between the limits 0 and 1, we find from (1.35) and (1.36), respectively,

$$(\beta^2 - \alpha^2)\int_0^1 xJ_0(\alpha x)J_0(\beta x)dx$$
$$= \alpha J_0'(\alpha)J_0(\beta) - \beta J_0'(\beta)J_0(\alpha), \quad (1.37)$$

$$\int_0^1 xJ_0^2(\alpha x)dx = \tfrac{1}{2}\{J_0^2(\alpha) + J_1^2(\alpha)\} \quad . \quad (1.38)$$

Corollary 1. If α, β ($\beta^2 \neq \alpha^2$) are two roots of the equation $J_0(x) = 0$, then

$$\int_0^1 xJ_0(\alpha x)J_0(\beta x)dx = 0.$$

This follows immediately from (1.37).

Corollary 2. If α, β ($\beta^2 \neq \alpha^2$) are two roots of the equation

$$xJ_0'(x) + HJ_0(x) = 0,$$

where H is a constant, then

$$\int_0^1 xJ_0(\alpha x)J_0(\beta x)dx = 0.$$

This also follows at once from (1.37).

Examples I

1. Show that

(i) $\int x^2 J_0(x)dx = x^2 J_1(x) + xJ_0(x) - \int J_0(x)dx$,

(ii) $\int x^3 J_0(x)dx = x(x^2 - 4)J_1(x) + 2x^2 J_0(x)$,

(iii) $\int x^4 J_0(x)dx = x^2(x^2 - 9)J_1(x) + 3x(x^2 - 3)J_0(x) + 9\int J_0(x)dx$.

(iv) $\int x \log x\, J_0(x)dx = J_0(x) + x \log x\, J_1(x)$.

2. Show that

$$\text{(i)}\quad \int_0^1 xJ_0(\alpha x)dx = \frac{1}{\alpha}J_1(\alpha),$$

$$\text{(ii)}\quad \int_0^1 x^2 J_0(\alpha x)dx = \frac{1}{\alpha}J_1(\alpha) + \frac{1}{\alpha^2}J_0(\alpha) - \frac{1}{\alpha^3}\int_0^\alpha J_0(t)dt,$$

$$\text{(iii)}\quad \int_0^1 x^3 J_0(\alpha x)dx = \frac{\alpha^2 - 4}{\alpha^3}J_1(\alpha) + \frac{2}{\alpha^2}J_0(\alpha),$$

$$\text{(iv)}\quad \int_0^1 x(1 - x^2)J_0(\alpha x)dx = \frac{4}{\alpha^3}J_1(\alpha) - \frac{2}{\alpha^2}J_0(\alpha).$$

3. If α is any root of the equation $J_0(x) = 0$, show that

$$\text{(i)}\quad \int_0^1 J_1(\alpha x)dx = \frac{1}{\alpha},$$

$$\text{(ii)}\quad \int_0^\alpha J_1(x)dx = 1,$$

$$\text{(iii)}\quad \int_0^\infty J_1(x)dx = 1.$$

If α ($\neq 0$) is a root of the equation $J_1(x) = 0$, show that

$$\text{(iv)}\quad \int_0^1 xJ_0(\alpha x)dx = 0.$$

4. If $J_0 \equiv J_0(x)$, $J_1 \equiv J_1(x)$, show that

(i) $\int J_0 J_1 dx = -\frac{1}{2}J_0^2$,

(ii) $\int xJ_0 J_1 dx = -\frac{1}{2}xJ_0^2 + \frac{1}{2}\int J_0^2 dx$,

(iii) $\int x^2 J_0 J_1 dx = \frac{1}{2}x^2 J_1^2$.

5. Show that

(i) $\int xJ_0^2 dx = \frac{1}{2}x^2(J_0^2 + J_1^2)$,

(ii) $\int xJ_1^2 dx = \frac{1}{2}x^2(J_0^2 + J_1^2) - xJ_0 J_1$,

(iii) $2\int x^2 J_0^2 dx = \frac{1}{3}x^3(J_0^2 + J_1^2) + \frac{1}{2}x^2 J_0 J_1 + \frac{1}{4}xJ_0^2 - \frac{1}{4}\int J_0^2 dx$,

(iv) $2\int x^2 J_1^2 dx = \frac{1}{3}x^3(J_0^2 + J_1^2) - \frac{3}{2}x^2 J_0 J_1 - \frac{3}{4}xJ_0^2 + \frac{3}{4}\int J_0^2 dx$,

(v) $3\int x^3 J_0^2 dx = \frac{1}{2}x^4(J_0^2 + J_1^2) + x^3 J_0 J_1 - x^2 J_1^2$,

(vi) $3\int x^3 J_1^2 dx = \frac{1}{2}x^4(J_0^2 + J_1^2) - 2x^3 J_0 J_1 + 2x^2 J_1^2$.

6. If $u = \int xJ_0(\alpha x)J_0(\beta x)dx$, $v = \int xJ_1(\alpha x)J_1(\beta x)dx$, show that

$$\alpha u - \beta v = xJ_1(\alpha x)J_0(\beta x),$$
$$\beta u - \alpha v = xJ_0(\alpha x)J_1(\beta x).$$

Deduce the values of u and v, and by differentiating partially with regard to β, and then putting $\beta = \alpha$, deduce the values of

$$\int x J_0^2(\alpha x)dx, \quad \int x J_1^2(\alpha x)dx.$$

7. If $f(x)$ is any Bessel function of zero order, show that

$$\frac{d}{dx}\frac{f(x)}{J_0(x)} = \frac{B}{xJ_0^2(x)}$$

where B is a constant; and hence that, if α, β are two consecutive positive roots of the equation $J_0(x) = 0$, the fraction $f(x)/J_0(x)$ increases steadily from $-\infty$ to $+\infty$ (or decreases steadily from $+\infty$ to $-\infty$) when x increases from α to β. Deduce that the equation $f(x) = 0$ has one root between α and β.

§ 12. *Behaviour of Bessel functions when x is large.*

If in Bessel's equation

$$x\frac{d^2y}{dx^2} + \frac{dy}{dx} + xy = 0,$$

we make the substitution

$$u = y\sqrt{x} \qquad . \quad . \quad . \quad (1.39)$$

we find that u satisfies the equation

$$\frac{d^2u}{dx^2} = -\left(1 + \frac{1}{4x^2}\right)u.$$

Now, when x is large enough, $1/4x^2$ is as small as we please compared with 1, and then we have approximately

$$\frac{d^2u}{dx^2} = -u$$

of which the general solution is $u = C\cos(x - \lambda)$, and we infer, by (1.39), that every solution of Bessel's equation behaves like

$$y = \frac{C\cos(x-\lambda)}{\sqrt{x}}$$

when x is large, where C, λ are constants.

In fact, it will be shown in § 80 that, if $x > 0$,

$$J_0(x) = \left(\frac{2}{\pi x}\right)^{\frac{1}{2}}\left\{\cos\left(x - \frac{\pi}{4}\right) + p(x)\right\}, \quad . \quad (1.40)$$

$$Y_0(x) = \left(\frac{2}{\pi x}\right)^{\frac{1}{2}}\left\{\sin\left(x - \frac{\pi}{4}\right) + q(x)\right\}, \quad . \quad (1.41)$$

where $p(x) \to 0$ and $q(x) \to 0$ when $x \to +\infty$.

Hence any solution $f(x)$ of Bessel's equation of zero order may be written in the form

$$\begin{aligned} f(x) &= \mathrm{AJ}_0(x) + \mathrm{BY}_0(x) \\ &= \left(\frac{2}{\pi x}\right)^{\frac{1}{2}}(\mathrm{A}^2 + \mathrm{B}^2)^{\frac{1}{2}}\{\cos(x - \lambda) + r(x)\}, \qquad (1.42) \end{aligned}$$

where $\lambda = \frac{1}{4}\pi + \tan^{-1}(\mathrm{B}/\mathrm{A})$, and $r(x) \to 0$ when $x \to +\infty$.

Corollary. Every solution tends to zero when $x \to +\infty$.

§ 13. *Roots of the equation $f(x) = 0$, where $f(x)$ denotes any Bessel function of zero order.*

If $f(x)$ denotes any Bessel function of zero order, it follows from (1.42) that, for large values of x, the roots of the equation $f(x) = 0$ are approximately those of $\cos(x - \lambda) = 0$. We infer that the equation $f(x) = 0$ has an infinite number of real roots, and that the large roots are $\lambda + (s - \frac{1}{2})\pi$, approximately, where s is any large positive integer.

In particular, the large positive roots of the equation $\mathrm{J}_0(x) = 0$ are $(s - \frac{1}{4})\pi$, those of $\mathrm{Y}_0(x) = 0$ are $(s - \frac{3}{4})\pi$, approximately, where s is any large positive integer.

§ 14. *None of these roots can be a repeated root.*

Proof. Let α be a root, and note, firstly, that $\alpha \neq 0$, since no Bessel function of zero order vanishes when $x = 0$.

Secondly, suppose that α could be a repeated root; then $f(\alpha) = 0$, $f'(\alpha) = 0$, and by substituting $x = \alpha$ in the differential equation

$$xf''(x) + f'(x) + xf(x) = 0, \qquad (1.43)$$

it would follow that $\alpha f''(\alpha) = 0$, and hence $f''(\alpha) = 0$, since $\alpha \neq 0$.

Moreover, by differentiating the differential equation and putting $x = \alpha$ again, it would follow that $f'''(\alpha) = 0$, and, by repeating this process, that all the derivatives of $f(x)$ would vanish when $x = \alpha$. Consequently, from Taylor's series

$$f(x) = f(\alpha) + (x - \alpha)f'(\alpha) + \frac{(x - \alpha)^2}{2!}f''(\alpha) + \ldots$$

we should have $f(x) \equiv 0$. Hence, $x = \alpha$ cannot be a repeated root.

§ 15. *Roots of the equations* $J_0(x) = 0$, $J_1(x) = 0$.

I. *The equation* $J_0(x) = 0$ *has an infinite number of real roots, all simple.*

This follows from §§ 13, 14 as a particular case. Another proof of this theorem will be given later (§ 96, II).

II. *The equation* $J_0(x) = 0$ *has no purely imaginary roots.*

Proof. Put $x = i\beta$, $(\beta \neq 0)$, in (1.1). Then

$$J_0(i\beta) = 1 + \frac{\beta^2}{2^2} + \frac{\beta^4}{2^2 . 4^2} + \frac{\beta^6}{2^2 . 4^2 . 6^2} + \dots$$

which cannot vanish, since all the terms on the right are positive. Hence $x = i\beta$ cannot be a root of $J_0(x) = 0$.

III. *The equation* $J_0(x) = 0$ *has no complex roots.*

Proof. Suppose that $a + ib$ could be a root $(a \neq 0, b \neq 0)$. Then the conjugate $a - ib$ would also be a root, because the coefficients in the series for $J_0(x)$ are all real; and, since $(a + ib)^2 \neq (a - ib)^2$, it would follow from § 11, Cor. 1, that

$$\int_0^1 xJ_0\{(a + ib)x\}J_0\{(a - ib)x\}dx = 0.$$

But this is impossible, because the integrand is positive throughout the range of integration, being the product of x and a conjugate pair of complex numbers. Hence, $a + ib$ cannot be a root.

IV. *The equation* $J_0'(x) = 0$ *has an infinite number of real roots.*

Proof. By Rolle's theorem, since $J_0(x)$ and $J_0'(x)$ are continuous, the equation $J_0'(x) = 0$ has at least one root between every pair of roots of $J_0(x) = 0$, and hence, by I, has an infinite number of real roots.

V. *The equation* $J_1(x) = 0$ *has an infinite number of real roots.*

This follows from IV, since $J_1(x) = - J_0'(x)$.

VI. *The equations* $J_0(x) = 0$, $J_1(x) = 0$ *have no common root.*

Proof. By I, $J_0(x)$, $J_0'(x)$ have no common root. But $J_1(x) = -J_0'(x)$, therefore $J_0(x)$, $J_1(x)$ have no common root.

VII. *The equation*

$$xJ_0'(x) + HJ_0(x) = 0, \quad . \quad . \quad (1.44)$$

where H is any real constant, has an infinite number of real roots.

Proof. Put

$$\phi(x) \equiv xJ_0'(x) + HJ_0(x), \quad . \quad . \quad (1.45)$$

and let α, β be a pair of consecutive positive roots of the equation $J_0(x) = 0$. Then

$$\phi(\alpha) = \alpha J_0'(\alpha), \quad \phi(\beta) = \beta J_0'(\beta). \quad . \quad (1.46)$$

Now, since α, β are simple roots of $J_0(x) = 0$, neither $J_0'(\alpha)$ nor $J_0'(\beta)$ can be zero ; and since $J_0(x)$ is continuous, $J_0'(\alpha)$ and $J_0'(\beta)$ must be of opposite sign. Since α and β are positive, it follows from (1.46) that $\phi(\alpha)$ and $\phi(\beta)$ are of opposite sign, and hence, since $\phi(x)$ is continuous, that the equation $\phi(x) = 0$ has a root between α and β. Consequently, the equation $\phi(x) = 0$ has an infinite number of real roots, at least one between every pair of consecutive roots of $J_0(x) = 0$.

§ 16. A few of the smaller positive roots of the equations $J_0(x) = 0$, $J_1(x) = 0$, along with those of the equations $J_n(x) = 0$, $(n = 2, 3, 4, 5)$ are given in a table at the beginning of the book.

§ 17. *Fourier-Bessel expansion of zero order.*

Let α_1, α_2, α_3 . . . denote the positive roots of the equation $J_0(x) = 0$, arranged in ascending order of magnitude.

In general (see § 99), any ordinary function of mathematical physics $f(x)$, arbitrarily defined in the interval

$0 < x < 1$, can be represented over this interval by an infinite series of the form

$$f(x) = A_1J_0(x\alpha_1) + A_2J_0(x\alpha_2) + A_3J_0(x\alpha_3) + \dots \quad (1.47)$$

which is called the *Fourier-Bessel expansion* of $f(x)$ of zero order.

For the expansion to hold good up to $x = 1$, a necessary condition is plainly $f(1) = 0$, since every term of the expansion vanishes when $x = 1$.

§ 18. Assuming the expansion to hold good, and that term-by-term integration can be justified, the coefficients $A_1, A_2, A_3 \dots$ can be determined with the aid of Lommel's integrals (§ 11).

Multiply (1.47) throughout by $xJ_0(x\alpha_s)dx$ and integrate between the limits 0 and 1; then the general term on the right-hand side will be

$$\int_0^1 xJ_0(x\alpha_r)J_0(x\alpha_s)dx$$

which vanishes when $r \neq s$, by § 11, Cor. 1. Consequently, every term on the right vanishes except the one in which $r = s$, and we get

$$\int_0^1 xf(x)J_0(x\alpha_s)dx = A_s\int_0^1 xJ_0^2(x\alpha_s)dx = \tfrac{1}{2}A_sJ_1^2(\alpha_s),$$

by (1.38), since $J_0(\alpha_s) = 0$. Hence

$$A_s = \frac{2}{J_1^2(\alpha_s)}\int_0^1 xf(x)J_0(x\alpha_s)dx. \quad (1.48)$$

The simplest cases in which this integral can be evaluated in terms of tabulated functions are

(i) $f(x) =$ a polynomial in x (see § 10), or $\log x$;
(ii) $f(x) = J_0(kx)$, where k is a constant (see § 11);
(iii) $f(x) = (1 - x^2)^p$, where $p > -1$ (see § 91).

Ex. Find the Fourier-Bessel expansion of $f(x) = 1 - x^2$.
In this case we have, by (1.48),

$$A_s = \frac{2}{J_1^2(\alpha_s)}\int_0^1 x(1 - x^2)J_0(x\alpha_s)dx,$$

and hence, replacing α by α_s in Exs. I, 2 (iv), and putting $J_0(\alpha_s) = 0$,

$$A_s = \frac{2}{J_1^2(\alpha_s)} \cdot \frac{4J_1(\alpha_s)}{\alpha_s^3} = \frac{8}{\alpha_s^3 J_1(\alpha_s)}.$$

Consequently,

$$1 - x^2 = 8\left\{\frac{J_0(x\alpha_1)}{\alpha_1^3 J_1(\alpha_1)} + \frac{J_0(x\alpha_2)}{\alpha_2^3 J_1(\alpha_2)} + \frac{J_0(x\alpha_3)}{\alpha_3^3 J_1(\alpha_3)} + \ldots\right\},$$

which may also be written

$$1 - x^2 = 8\sum_{\alpha} \frac{J_0(\alpha x)}{\alpha^3 J_1(\alpha)} \quad . \quad . \quad . \quad (1.49)$$

where the summation extends over the positive roots of the equation $J_0(x) = 0$.

This expansion holds good over the range $-1 < x < 1$, since $1 - x^2$ is even.

We may gain an idea of the numerical values of the coefficients from the table at the beginning of the book; thus we find approximately

$$1 - x^2 = 1{\cdot}108J_0(x\alpha_1) - {\cdot}140J_0(x\alpha_2) + {\cdot}045J_0(x\alpha_3) - {\cdot}021J_0(x\alpha_4) + {\cdot}012J_0(x\alpha_5) - \ldots$$

§ 19. *Dini expansion of zero order.*

An expansion similar to (1.47), but based upon the roots of the equation

$$xJ_0'(x) + HJ_0(x) = 0, \quad . \quad . \quad (1.50)$$

is called * the *Dini expansion* of $f(x)$ of zero order. Three cases may be distinguished, depending upon the values of the constant H.

I. If $H > 0$, the Dini expansion has exactly the same form as (1.47), viz.

$$f(x) = A_1J_0(x\alpha_1) + A_2J_0(x\alpha_2) + A_3J_0(x\alpha_3) + \ldots$$

where $\alpha_1, \alpha_2, \alpha_3, \ldots$ are the positive roots of (1.50), and $0 < x < 1$.

The coefficients are determined in the same manner as before. We multiply both sides by $xJ_0(x\alpha_s)dx$, and integrate between the limits 0 and 1; then the general term on the right-hand side will be

$$\int_0^1 xJ_0(x\alpha_r)J_0(x\alpha_s)dx$$

* Watson, p. 580.

which vanishes when $r \neq s$, by § 11, Cor. 2. Consequently every term on the right vanishes except the one in which $r = s$, and we get

$$\int_0^1 xf(x)J_0(x\alpha_s)dx = A_s\int_0^1 xJ_0^2(x\alpha_s)dx = \tfrac{1}{2}A_s\{J_0^2(\alpha_s) + J_1^2(\alpha_s)\},$$

by (1.38), which determines the coefficient A_s.

II. If $H = 0$, equation (1.50) becomes

$$xJ_0'(x) = 0 \quad \text{or} \quad xJ_1(x) = 0,$$

which has a double root $x = 0$, and in this case the series has an initial constant term, thus

$$f(x) = A_0 + A_1J_0(x\alpha_1) + A_2J_0(x\alpha_2) + \ldots \qquad (1.51)$$

which may be regarded as an expansion based upon the roots, $0, \alpha_1, \alpha_2, \alpha_3 \ldots$ of the equation $J_1(x) = 0$. The constant A_0 may be obtained by multiplying throughout by $x\,dx$ and integrating from 0 to 1. Thus we get, using Exs. I, 3 (iv),

$$\int_0^1 xf(x)dx = A_0\int_0^1 x\,dx = \tfrac{1}{2}A_0$$

which gives A_0.

Since $J_1(\alpha_s) = 0$, the constants A_s $(s \neq 0)$ are given now by

$$\int_0^1 xf(x)J_0(x\alpha_s)dx = \tfrac{1}{2}A_sJ_0^2(\alpha_s).$$

III. If $H < 0$, the equation (1.50) has two purely imaginary roots, and the Dini expansion involves an initial term depending on them (see § 98, III).

Examples II

1. If α is a typical positive root of $J_0(x) = 0$, obtain the following expansions :—

(i) $$1 = 2\sum_\alpha \frac{J_0(\alpha x)}{\alpha J_1(\alpha)}.$$

(ii) $$x^2 = 2\sum_\alpha \frac{\alpha^2 - 4}{\alpha^3 J_1(\alpha)}J_0(\alpha x).$$

(iii) $J_0(kx) = 2J_0(k)\sum_\alpha \frac{\alpha J_0(\alpha x)}{(\alpha^2 - k^2)J_1(\alpha)}$.

(iv) $x = 2\sum_\alpha \left\{\frac{1}{\alpha J_1(\alpha)} - \frac{1}{\alpha^3 J_1^2(\alpha)}\int_0^\alpha J_0(t)dt\right\} J_0(\alpha x)$.

(v) $\log \frac{1}{x} = 2\sum_\alpha \frac{J_0(\alpha x)}{\alpha^2 J_1^2(\alpha)}$.

2. If α denotes a typical positive root of $J_1(x) = 0$, obtain the following expansions :—

$$\text{(i)}\ x^2 = \frac{1}{2} + 4\sum_\alpha \frac{J_0(\alpha x)}{\alpha^2 J_0(\alpha)}.$$

$$\text{(ii)}\ (1 - x^2)^2 = \frac{1}{3} - 64\sum_\alpha \frac{J_0(\alpha x)}{\alpha^4 J_0(\alpha)}.$$

3. In an expansion of the form

$$f(x) = A_0 + \sum_\alpha AJ_0(\alpha x), \quad (0 < x < 1),$$

where $J_1(\alpha) = 0$, show that $\frac{1}{2}A_0$ is the average value of $xf(x)$ over the interval $0 < x < 1$.

4. If α is a typical positive root of $J_0(x) = 0$, and $-a < r < a$, show that

$$a^2 - r^2 = 8a^2\sum_\alpha \frac{1}{\alpha^3 J_1(\alpha)} J_0\left(\frac{\alpha r}{a}\right).$$

5. If $f(x)$ is defined arbitrarily in the interval $0 < x < h$, and α is a typical positive root of $J_0(x) = 0$, show that

$$f(x) = \sum_\alpha \frac{2J_0\left(\frac{\alpha x}{h}\right)}{h^2 J_1^2(\alpha)}\int_0^h f(t)J_0\left(\frac{\alpha t}{h}\right) t\, dt.$$

CHAPTER II

APPLICATIONS

§ 20. *Uniformly stretched uniform membranes.*

Bessel functions find their simplest applications in certain ideal problems of mathematical physics, e.g. the vibrations of a uniformly stretched uniform circular membrane.

It is assumed that the ideal membrane is perfectly flexible, i.e. that the stress across any line drawn on the

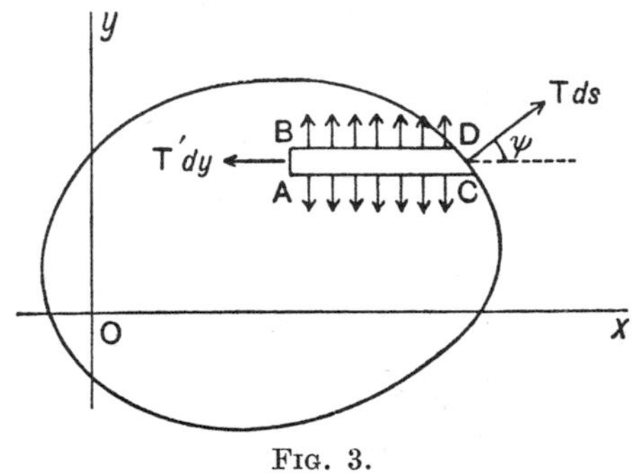

Fig. 3.

membrane is a tension perpendicular to the line at every point and in the tangent plane to the membrane.

Consider such a membrane, plane in its equilibrium position, under the action of a uniform tension T per unit length over its boundary, and let T′ be the tension per unit length across a line element AB (Fig. 3). Take rectangular axes in the plane of the membrane, with Oy parallel to AB, and let AB $= dy$. Through A, B draw lines parallel to Ox, meeting the boundary in C, D respectively. Let CD $= ds$

and let the normal to the element CD make an angle ψ with Ox. Consider the equilibrium of the element of membrane ACDB. Resolving in the direction of the x-axis, and remembering that the stress is everywhere perpendicular to the boundary, so that the stresses across the edges AC, BD have no components in the direction of the x-axis, we have

$$T'dy = Tds \,.\, \cos\psi = T \,.\, ds\cos\psi = Tdy,$$

and hence, dividing by dy,

$$T' = T\,;$$

thus the tension per unit length across any line drawn in the membrane is constant and equal to that over the boundary.

§ 21. *Differential equation of the small vibrations of such a membrane.*

Suppose the plane of the membrane horizontal and the effect of gravity negligible, and consider a motion in which the displacement of every point is small and the gradient is everywhere small. Let ABCD (Fig. 4) be a rectangular element of edges dx, dy, with its centre at the point (x, y) and its edges parallel to the co-ordinate axes; and let z be the displacement of the centre from its equilibrium position.

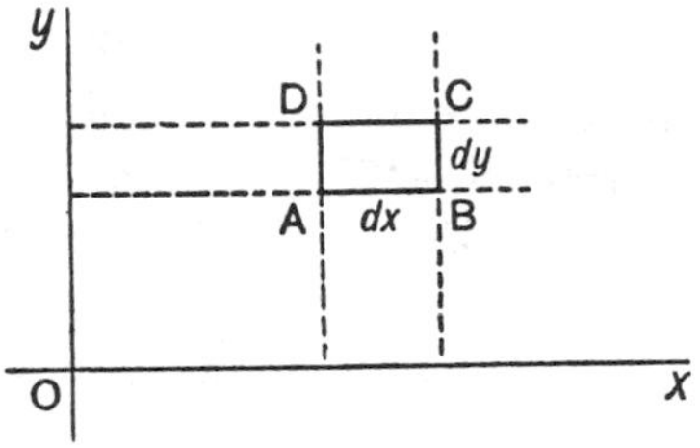

Fig. 4.

The motion of the element is caused by the tensions acting across its boundary and by possible external forces. Now the tension across a line element of length dy passing through the centre parallel to the edges AD, BC is Tdy, and its component perpendicular to the xy plane is approximately

$$Tdy \,.\, \frac{\partial z}{\partial x}, = T\frac{\partial z}{\partial x}dy,$$

since the gradient is small. The corresponding components across the edges BC, AD are respectively

$$\mathrm{T}\frac{\partial z}{\partial x}dy \pm \frac{\partial}{\partial x}\left(\mathrm{T}\frac{\partial z}{\partial x}dy\right)\frac{dx}{2}.$$

These act in opposite directions on the elementary rectangle, and together they contribute

$$\mathrm{T}\frac{\partial^2 z}{\partial x^2} \,.\, dx\,dy$$

to the forces causing its motion. A similar contribution

$$\mathrm{T}\frac{\partial^2 z}{\partial y^2} \,.\, dx\,dy$$

is supplied by the tensions across the edges AB, DC. If we suppose that in addition there is an external force Z per unit area, and that the mass per unit area is σ, the equation of motion of the element is

$$\sigma\, dx\, dy \,.\, \frac{\partial^2 z}{\partial t^2} = \mathrm{T}\left(\frac{\partial^2 z}{\partial x^2} + \frac{\partial^2 z}{\partial y^2}\right) . \, dx\, dy + \mathrm{Z} \,.\, dx\, dy + \ldots$$

the dots at the end indicating unknown terms of higher order of smallness in dx, dy. Hence, dividing by $dx\,dy$ and putting $c^2 = \mathrm{T}/\sigma$, we obtain, in the limit, the differential equation

$$\frac{\partial^2 z}{\partial t^2} = c^2\left(\frac{\partial^2 z}{\partial x^2} + \frac{\partial^2 z}{\partial y^2}\right) + \frac{\mathrm{Z}}{\sigma}. \qquad (2.1)$$

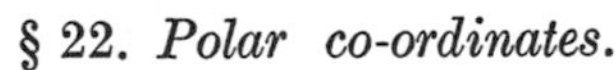

§ 22. *Polar co-ordinates.*

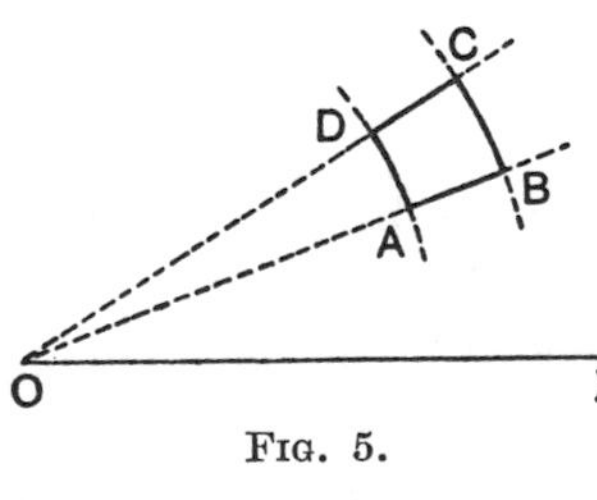

Fig. 5.

In polar co-ordinates, if we consider the motion of the element ABCD (Fig. 5) bounded by the circles $r \pm \frac{1}{2}dr$ and the radius vectors $\theta \pm \frac{1}{2}d\theta$, the tensions across the edges AD, BC contribute a component

$$\frac{\partial}{\partial r}\left(\mathrm{T}\frac{\partial z}{\partial r} . \, r\, d\theta\right) . \, dr, \quad = \mathrm{T}\left(\frac{\partial^2 z}{\partial r^2} + \frac{1}{r}\frac{\partial z}{\partial r}\right) . \, r\, dr\, d\theta,$$

to the forces causing the motion of the element, and the contribution from the edges AB, DC is

$$\frac{\partial}{\partial\theta}\left(\mathrm{T}\frac{\partial z}{r\partial\theta} \cdot dr\right) . d\theta, = \mathrm{T}\left(\frac{1}{r^2}\frac{\partial^2 z}{\partial\theta^2}\right) . r\,dr\,d\theta.$$

The equation of motion of the element now leads to the differential equation

$$\frac{\partial^2 z}{\partial t^2} = c^2\left(\frac{\partial^2 z}{\partial r^2} + \frac{1}{r}\frac{\partial z}{\partial r} + \frac{1}{r^2}\frac{\partial^2 z}{\partial\theta^2}\right) + \frac{\mathrm{Z}}{\sigma}. \qquad (2.2)$$

This could also be deduced analytically from (2.1) by changing the independent variables from x, y to r, θ.

§ 23. *Special cases.*

I. If the membrane is vibrating freely, i.e. if there is no external force, then $\mathrm{Z} = 0$. If, in addition, the motion is symmetrical about the origin, i.e. z is independent of θ, we have $\partial z/\partial\theta = 0$, and under these conditions equation (2.2) reduces to

$$\frac{\partial^2 z}{\partial t^2} = c^2\left(\frac{\partial^2 z}{\partial r^2} + \frac{1}{r}\frac{\partial z}{\partial r}\right). \qquad (2.3)$$

II. If z is the same at every point along any line parallel to the y-axis, so that z is independent of y, then $\partial z/\partial y = 0$ and equation (2.1) reduces to

$$\frac{\partial^2 z}{\partial t^2} = c^2\frac{\partial^2 z}{\partial x^2}. \qquad (2.4)$$

The general solution of this equation is known to be

$$z = f(x - ct) + \mathrm{F}(x + ct), \qquad (2{\cdot}5)$$

where f, F are arbitrary functions, and by considering either term of this solution we are led to a physical meaning of the constant c. Thus, if we consider the first term and put

$$z_1 = f(x - ct), \qquad (2.6)$$

then z_1 represents a displacement in which, at time $t = 0$,

$$z_1 = f(x). \qquad (2{\cdot}7)$$

At time t' later, taking a new origin at the point $x = ct'$, $z_1 = 0$, and putting $x = x' + ct'$, we find

$$z_1 = f(x').$$

Comparing this with (2.7), we infer that (2.6) represents a displacement which is travelling in the positive direction of the x-axis, unchanged in shape, and covering a distance ct' in time t', so that the displacement is travelling with uniform speed c.

Similarly, if we put

$$z_2 = \mathrm{F}(x + ct)$$

we should find that z_2 represents a displacement travelling in the negative direction of the x-axis with uniform speed c.

Thus, c is the speed with which a one-dimensional displacement could travel, unchanging in shape, across the membrane. It may be verified that $c = \sqrt{(\mathrm{T}/\sigma)}$, has the dimensions of velocity.

§ 24. *Normal modes of vibration.*

It is known that, when any mechanical system is vibrating freely about a position of stable equilibrium, it has *normal modes of vibration.*

A normal mode of vibration is one in which all the particles of the system vibrate with the same period and pass through their mean positions simultaneously.

§ 25. *Symmetrical normal modes of vibration of a circular membrane, with the circumference fixed.*

Let a be the radius of the membrane. The motion being symmetrical about the centre, the equation to be satisfied by the displacement z is (2.3), viz.

$$\frac{\partial^2 z}{\partial t^2} = c^2\left(\frac{\partial^2 z}{\partial r^2} + \frac{1}{r}\frac{\partial z}{\partial r}\right). \qquad (2.8)$$

For the normal modes of vibration, z must be of the form

$$z = \mathrm{R}\cos(\omega t - \epsilon), \qquad (2.9)$$

where R depends upon r only; and since the circumference is fixed, the solution must satisfy the condition

$$z = 0 \text{ when } r = a, \text{ for all values of } t. \qquad (2.10)$$

Substituting (2.9) in (2.8) and dividing throughout by $\cos(\omega t - \epsilon)$, we find that R must satisfy the equation

$$\frac{d^2\mathrm{R}}{dr^2} + \frac{1}{r}\frac{d\mathrm{R}}{dr} + \frac{\omega^2}{c^2}\mathrm{R} = 0,$$

and hence, by § 7, that R must be of the form

$$R = AJ_0\left(\frac{\omega r}{c}\right) + BY_0\left(\frac{\omega r}{c}\right).$$

Now, in the present problem, z is small at the centre $r = 0$. But the Bessel function of the second kind $Y_0(\omega r/c)$ becomes infinite at $r = 0$ (§ 6). Consequently, the constant B must be zero, and hence

$$z = AJ_0\left(\frac{\omega r}{c}\right) \cos (\omega t - \epsilon). \quad . \quad . \quad (2.11)$$

It remains to satisfy the boundary condition (2.10). Substituting $z = 0$, $r = a$, we get

$$J_0\left(\frac{\omega a}{c}\right) = 0. \quad . \quad . \quad . \quad (2.12)$$

This equation determines the possible values of ω and gives, since ω must be positive,

$$\frac{\omega a}{c} = \alpha_1, \alpha_2, \alpha_3, \ . \ . \ . \ \alpha_s \ . \ . \ .$$

where α_1, α_2, . . . are the positive roots of the equation $J_0(x) = 0$. It follows that, for the ideal membrane, there are an infinite number of normal modes of vibration, whose periods $2\pi/\omega_1$, $2\pi/\omega_2$, . . . $2\pi/\omega_s$, . . . are determined by

$$\omega_1 = \frac{c\alpha_1}{a}, \quad \omega_2 = \frac{c\alpha_2}{a}, \ . \ . \ . \ \omega_s = \frac{c\alpha_s}{a}, \ . \ . \ . \ . \quad (2.13)$$

If we distinguish the successive normal modes by the suffixes 1, 2, 3, . . . we can write

$$z_1 = C_1J_0\left(\frac{r\alpha_1}{a}\right) \cos (\omega_1 t - \epsilon_1),$$

$$z_2 = C_2J_0\left(\frac{r\alpha_2}{a}\right) \cos (\omega_2 t - \epsilon_2),$$

$$z_3 = C_3J_0\left(\frac{r\alpha_3}{a}\right) \cos (\omega_3 t - \epsilon_3), \ . \ . \ .$$

where C_1, C_2, C_3, . . . denote arbitrary constants which must, however, be small compared with a, so as to keep the displacement small at every point of the membrane.

In the first normal mode a radial section through the membrane at any instant has the shape of the graph of $y = J_0(x)$ from $x = 0$ to $x = \alpha_1$, since r varies from 0 to a (Fig. 6.1).

In the second mode the shape is that of $y = J_0(x)$ from $x = 0$ to $x = \alpha_2$. There is a nodal circle at $r = a\alpha_1/\alpha_2$ (Fig. 6.2).

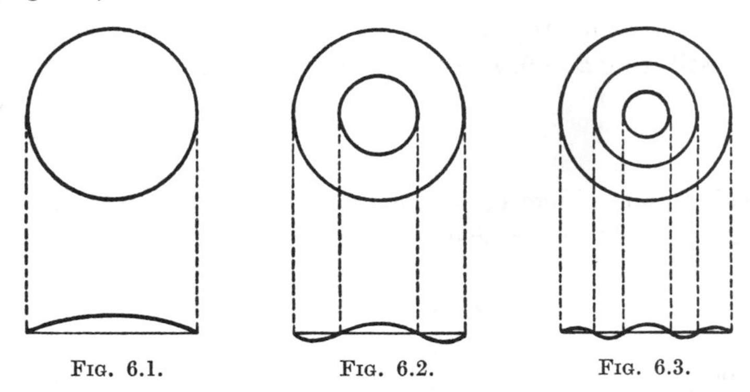

Fig. 6.1. Fig. 6.2. Fig. 6.3.

In the third mode the shape is that of $y = J_0(x)$ from $x = 0$ to $x = \alpha_3$. There are two nodal circles at $r = a\alpha_1/\alpha_3$, $r = a\alpha_2/\alpha_3$, respectively (Fig. 6.3); and so on.

§ 26. *General initial conditions.*

The most general motion of the membrane can be represented as a sum of arbitrary multiples of the normal modes, thus

$$z = \sum_{s=1}^{\infty} C_s J_0\left(\frac{r\alpha_s}{a}\right) \cos(\omega_s t - \epsilon_s), \quad . \quad (2.14)$$

or, what amounts to the same thing,

$$z = \sum_{s=1}^{\infty} J_0\left(\frac{r\alpha_s}{a}\right)(A_s \cos \omega_s t + B_s \sin \omega_s t). \quad (2.15)$$

Every term of this expression satisfies the differential equation (2.8) and the condition that $z = 0$ at $r = a$; and the coefficients A_s, B_s can be chosen so as to satisfy arbitrary initial conditions. For, let these be

$$z = \phi(r), \quad 0 < r < a, \quad t = 0; \qquad . \quad \text{(i)}$$

$$\partial z/\partial t = \psi(r), \quad 0 < r < a, \quad t = 0. \qquad . \quad \text{(ii)}$$

Then, putting $t = 0$ in (2.15) and using (i), we get

$$\phi(r) = A_1 J_0\left(\frac{r\alpha_1}{a}\right) + A_2 J_0\left(\frac{r\alpha_2}{a}\right) + A_3 J_0\left(\frac{r\alpha_3}{a}\right) + \ldots$$

Also, differentiating (2.15) with respect to t and then putting $t = 0$, and using (ii), we get

$$\psi(r) = \omega_1 B_1 J_0\left(\frac{r\alpha_1}{a}\right) + \omega_2 B_2 J_0\left(\frac{r\alpha_2}{a}\right) + \omega_3 B_3 J_0\left(\frac{r\alpha_3}{a}\right) + \ldots$$

The coefficients A_s, B_s have to be chosen so that these equations are satisfied by values of r between $r = 0$ and $r = a$. If we put $x = r/a$, the equations become

$$\phi(ax) = A_1 J_0(x\alpha_1) + A_2 J_0(x\alpha_2) + A_3 J_0(x\alpha_3) + \ldots$$

$$\psi(ax) = \omega_1 B_1 J_0(x\alpha_1) + \omega_2 B_2 J_0(x\alpha_2) + \omega_3 B_3 J_0 x\alpha_3) + \ldots$$

which have to be satisfied between $x = 0$ and $x = 1$. Consequently, by (1.48),

$$A_s = \frac{2}{J_1^2(\alpha_s)}\int_0^1 x\phi(ax) J_0(x\alpha_s)dx, \quad . \qquad . \quad (2.16)$$

$$B_s = \frac{2}{\omega_s J_1^2(\alpha_s)}\int_0^1 x\psi(ax) J_0(x\alpha_s)dx. \qquad . \quad (2.17)$$

With these values of the coefficients, the value of z is given by (2.15).

EXAMPLES III

1. If $z = C(a^2 - r^2)$ and $\partial z/\partial t = 0$ when $t = 0$, show that

$$z = 8\,Ca^2\left\{\frac{J_0\left(\frac{r\alpha_1}{a}\right)\cos\frac{c\alpha_1 t}{a}}{\alpha_1^3 J_1(\alpha_1)} + \frac{J_0\left(\frac{r\alpha_2}{a}\right)\cos\frac{c\alpha_2 t}{a}}{\alpha_2^3 J_1(\alpha_2)} + \ldots\right\}$$

2. A uniformly stretched circular membrane, with the circumference fixed, is acted upon by a force $Z = F \cos pt$ per unit area,

where F is a constant. Show that, in the notation used above, the forced oscillation thus caused is given by

$$z = \frac{\mathrm{F}}{\sigma p^2 \mathrm{J}_0(pa/c)} \left\{ \mathrm{J}_0\left(\frac{pr}{c}\right) - \mathrm{J}_0\left(\frac{pa}{c}\right) \right\} \cos pt$$

provided $2\pi/p$ is not a natural period of vibration.

3. If the membrane has the form of a circular annulus of radii a, b, the condition $z = 0$ being satisfied round both the circles $r = a$, $r = b$, show that the periods of the normal modes of vibration are $2\pi/\omega$ where ω satisfies the equation

$$\mathrm{J}_0\left(\frac{\omega a}{c}\right)\mathrm{Y}_0\left(\frac{\omega b}{c}\right) - \mathrm{J}_0\left(\frac{\omega b}{c}\right)\mathrm{Y}_0\left(\frac{\omega a}{c}\right) = 0.$$

If $b > a$, show that the large roots of this equation are given by $\omega(b - a)/c = s\pi$, approximately, where s is a large positive integer. [Use (1.40), (1.41).]

§ 27. *Small oscillations of a uniform flexible hanging chain, in a vertical plane.*

Take the origin O at the equilibrium position of the lower end of the chain (Fig. 7). Let l be its length, λ its mass per unit length.

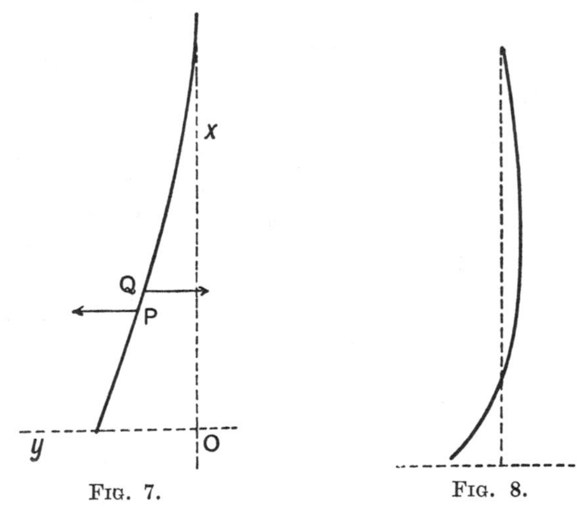

Fig. 7. Fig. 8.

Consider the motion of an element PQ of length dx, with its midpoint at a height x above O. Let T be the tension at the middle point of the element. The horizontal component of this tension is approximately $-\mathrm{T}\partial y/\partial x$, and the corresponding components at the ends P, Q are

$$-\mathrm{T}\frac{\partial y}{\partial x} \pm \frac{\partial}{\partial x}\left(-\mathrm{T}\frac{\partial y}{\partial x}\right)\frac{dx}{2}.$$

These act in opposite directions on the element and cause its motion ; the equation of motion is therefore

$$\lambda\, dx \,.\, \frac{\partial^2 y}{\partial t^2} = \frac{\partial}{\partial x}\left(\mathrm{T}\frac{\partial y}{\partial x}\right) \,.\, dx + \ldots$$

the dots at the end indicating unknown terms of higher order of smallness.

Now $\mathrm{T} = g\lambda x$ approximately, the vertical motion being ignored ; hence

$$\lambda\, dx \,.\, \frac{\partial^2 y}{\partial t^2} = \frac{\partial}{\partial x}\left(g\lambda x\frac{\partial y}{\partial x}\right)dx + \ldots$$

Dividing by $\lambda\, dx$, since λ is constant, we have, in the limit, the differential equation

$$\frac{\partial^2 y}{\partial t^2} = g\frac{\partial}{\partial x}\left(x\frac{\partial y}{\partial x}\right). \quad . \quad . \quad . \quad (2.18)$$

§ 28. To find the normal modes of vibration, we make the substitution

$$y = \mathrm{X}\cos(\omega t - \epsilon), \quad . \quad . \quad (2.19)$$

where X is a function of x only, and after dividing by $\cos(\omega t - \epsilon)$ we find that X must satisfy the equation

$$x\frac{d^2\mathrm{X}}{dx^2} + \frac{d\mathrm{X}}{dx} + \frac{\omega^2}{g}\,\mathrm{X} = 0. \quad . \quad . \quad (2.20).$$

This, as it stands, is not a Bessel equation, but the substitution *

$$x = \tfrac{1}{4}g\tau^2 \quad . \quad . \quad . \quad (2.21)$$

* See Lamb : "Higher Mechanics," p. 219, for the physical meaning of the new independent variable τ.

transforms it into the Bessel equation

$$\frac{d^2X}{d\tau^2} + \frac{1}{\tau}\frac{dX}{d\tau} + \omega^2 X = 0, \qquad (2.22)$$

from which follows

$$X = AJ_0(\omega\tau) + BY_0(\omega\tau). \qquad (2.23)$$

The constant B must be zero, for the same kind of reason as in § 25, and hence

$$y = AJ_0(\omega\tau)\cos(\omega t - \epsilon)$$
$$= AJ_0\left(2\omega\sqrt{\frac{x}{g}}\right)\cos(\omega t - \epsilon). \qquad (2.24)$$

The condition that $y = 0$ when $x = l$ gives the equation

$$J_0\left(2\omega\sqrt{\frac{l}{g}}\right) = 0, \qquad (2.25)$$

which determines the values of ω, and hence the periods of the normal modes of vibration ; thus

$$2\omega\sqrt{\frac{l}{g}} = \alpha_1, \alpha_2, \ldots \alpha_s, \ldots \qquad (2.26)$$

where $\alpha_1, \alpha_2, \ldots$ are the positive roots of $J_0(x) = 0$, and the corresponding periods can be written

$$\frac{2}{\alpha_1} \times 2\pi\sqrt{\frac{l}{g}}, \ \frac{2}{\alpha_2} \times 2\pi\sqrt{\frac{l}{g}}, \ \ldots \ \frac{2}{\alpha_s} \times 2\pi\sqrt{\frac{l}{g}}, \ \ldots$$

or

$$0{\cdot}832 \times 2\pi\sqrt{\frac{l}{g}}, \ 0{\cdot}363 \times 2\pi\sqrt{\frac{l}{g}}, \ \ldots$$

The corresponding normal modes are of the form

$$y_1 = C_1J_0\left(\alpha_1\sqrt{\frac{x}{l}}\right)\cos(\omega_1 t - \epsilon_1),$$

$$y_2 = C_2J_0\left(\alpha_2\sqrt{\frac{x}{l}}\right)\cos(\omega_2 t - \epsilon_2), \ \ldots$$

In the second normal mode (Fig. 8) there is a node given by $\alpha_2\sqrt{(x/l)} = \alpha_1$, or

$$\frac{x}{l} = \left(\frac{\alpha_1}{\alpha_2}\right)^2 = 0{\cdot}190.$$

The part of the chain below the node vibrates in its own first normal mode, i.e. the period of the second mode for the whole chain is that of the first mode for the part below the node.

In the third mode there are two nodes, in the fourth three nodes, and so on.

Examples IV

1. If the initial conditions are

$$\begin{aligned} y &= m(l - x), & 0 < x < l, \quad t = 0, \\ \partial y/\partial t &= 0, & 0 < x < l, \quad t = 0, \end{aligned}$$

show that

$$y = 8ml\left\{J_0\left(\alpha_1\sqrt{\frac{x}{l}}\right)\frac{\cos\omega_1 t}{\alpha_1^3 J_1(\alpha_1)} + J_0\left(\alpha_2\sqrt{\frac{x}{l}}\right)\frac{\cos\omega_2 t}{\alpha_2^3 J_1(\alpha_2)} + \ldots\right\},$$

where $\omega_s = \frac{1}{2}\alpha_s\sqrt{(g/l)}$.

2. Show how to satisfy the general initial conditions

$$\begin{aligned} y &= \phi(x), & 0 < x < l, \quad t = 0, \\ \partial y/\partial t &= \psi(x), & 0 < x < l, \quad t = 0. \end{aligned}$$

[Cf. § 26.]

3. A uniform flexible chain of length l, suspended from one end, rotates about a vertical axis through that end, in relative equilibrium. Show that the possible values of ω, the angular velocity, are given by

$$\omega_1 = \tfrac{1}{2}\alpha_1\sqrt{(g/l)}, \quad \omega_2 = \tfrac{1}{2}\alpha_2\sqrt{(g/l)}, \quad \omega_3 = \tfrac{1}{2}\alpha_3\sqrt{(g/l)}, \ \ldots$$

where $\alpha_1, \alpha_2, \alpha_3, \ldots$ are the positive roots of $J_0(x) = 0$; and hence that the periods of rotation are the same as the periods of the normal modes of vibration in a vertical plane.

§ 29. *Conduction of heat in an isotropic solid.*

Suppose that one face of a uniform plate, of thickness d, is maintained at temperature u and the opposite face at temperature u_1. Also, suppose the plate to be intersected at right angles by a cylindrical surface of cross-sectional area S. Then the quantity of heat Q that flows perpendicularly to the faces from the first to the second in time t within the cylindrical surface varies directly as $u - u_1$, S and t and inversely as d, and we write

$$Q = K \cdot \frac{u - u_1}{d} \cdot S \cdot t, \qquad (2.27)$$

where K is a constant, called the *coefficient of thermal conductivity* of the material of which the plate is made. This is the law, founded on experiment, upon which the classical theory of the conduction of heat is based.

§ 30. In order to adapt the law to an isotropic body bounded by any surface, we imagine the body dissected by isothermal surfaces. An isothermal surface is one at every point of which the temperature is momentarily the same. The direction of flow of heat at any point is normal to the isothermal surface through the point, because there is no temperature-gradient in any direction tangential to the surface.

Let u and $u + du$ be the temperatures of two neighbouring isothermal surfaces, and $d\text{S}$ an element of the surface u at a point where dn is the normal distance between it and the surface $u + du$. Then the quantity of heat $d\text{Q}$ that passes in time dt across $d\text{S}$ in the direction from the surface u to the surface $u + du$ is given by

$$d\text{Q} = -\text{K}\frac{\partial u}{\partial n}d\text{S}\,dt. \qquad (2.28)$$

Next, let $d\text{S}$ be any element of surface at any point of the body, and let the normal to $d\text{S}$ make an angle θ with the normal to the isothermal through that point (Fig. 9).

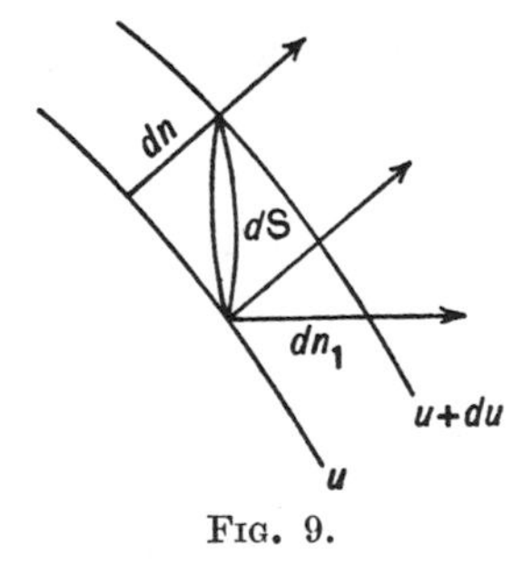

FIG. 9.

Let u be the temperature of the isothermal, and let dn be the element of the normal to the isothermal between it and the neighbouring isothermal $u + du$. Also, let dn_1 be the element of the normal to $d\text{S}$ between the same two isothermals. Then, since the projection of $d\text{S}$ on the isothermal surface is $d\text{S} \cos \theta$, the quantity of heat $d\text{Q}$ that flows across $d\text{S}$ in time dt is given by

$$d\text{Q} = -\text{K}\frac{\partial u}{\partial n} \,.\, d\text{S} \cos \theta \,.\, dt.$$

But $dn = dn_1 \cos\theta$, and therefore

$$dQ = -K\frac{\partial u}{\partial n_1}dS\,dt. \quad . \quad . \quad . \quad (2.29)$$

Hence, since $\partial u/\partial n_1$ is the temperature gradient perpendicular to dS, the law (2.28) holds good for any element of surface dS, whether it is part of an isothermal or not.

§ 31. *Differential equation of the conduction of heat.*

To find the differential equation satisfied by the temperature at any point in the interior of an isotropic body, we begin by considering a small rectangular parallelepiped with its centre at the point $P(x, y, z)$ and its edges, of lengths dx, dy, dz, parallel to a convenient set of Cartesian axes (Fig. 10).

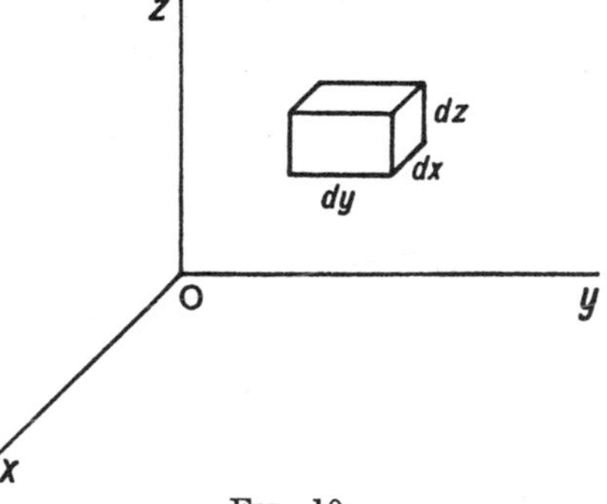

FIG. 10.

Let u be the temperature at P, and let $\partial u/\partial x$, $\partial u/\partial y$, $\partial u/\partial z$, $\partial u/\partial t$ be its rates of change with respect to the space co-ordinates x, y, z and the time t. We shall find two expressions for the increase in the quantity of heat contained within the parallelepiped in time dt, and equate them.

Firstly, the quantity of heat that flows across a section through P parallel to the yz plane is, by (2.28),

$$-K\frac{\partial u}{\partial x}\,.\,dy\,dz\,.\,dt.$$

The quantities that flow across the two faces parallel to this section are therefore

$$\left(-K\frac{\partial u}{\partial x} \pm K\frac{\partial^2 u}{\partial x^2}\frac{dx}{2}\right).\,dy\,dz\,.\,dt.$$

The upper sign corresponds to an inflow across the face nearer to the yz plane, the lower to an outflow across the opposite face, and by subtraction the nett inflow across the two faces is

$$K\frac{\partial^2 u}{\partial x^2}dx\,dy\,dz\,dt.$$

There are similar contributions to the inflow across the other two pairs of opposite faces, and by adding it follows that a first expression for the increment dQ in the quantity of heat contained within the parallelepiped is given by

$$dQ = K\left(\frac{\partial^2 u}{\partial x^2} + \frac{\partial^2 u}{\partial y^2} + \frac{\partial^2 u}{\partial z^2}\right)dx\,dy\,dz\,dt. \quad . \quad (2.30)$$

Secondly, let ρ be the density of the material and s its specific heat. Then the mass of the parallelepiped is $\rho\,dx\,dy\,dz$, and since the increase in temperature during the interval dt is $\partial u/\partial t\,.\,dt$, a second expression for the increment dQ is given by

$$dQ = s\,.\,\rho\,dx\,dy\,dz\,.\,\frac{\partial u}{\partial t}dt. \quad . \quad . \quad (2.31)$$

Equating these two expressions for dQ, and dividing by $s\rho\,dx\,dy\,dz\,dt$, we find the equation satisfied by the temperature u, viz.

$$\frac{\partial u}{\partial t} = \kappa\left(\frac{\partial^2 u}{\partial x^2} + \frac{\partial^2 u}{\partial y^2} + \frac{\partial^2 u}{\partial z^2}\right), \quad . \quad . \quad (2.32)$$

where
$$\kappa = \frac{K}{s\rho}. \quad . \quad . \quad . \quad (2.33)$$

§ 32. *Special cases.*

I. If the flow is *two-dimensional* and parallel to the xy plane, so that u is independent of z and $\partial u/\partial z = 0$, the equation reduces to

$$\frac{\partial u}{\partial t} = \kappa\left(\frac{\partial^2 u}{\partial x^2} + \frac{\partial^2 u}{\partial y^2}\right). \quad . \quad . \quad (2.34)$$

In polar co-ordinates (cf. § 22), the same equation reads

$$\frac{\partial u}{\partial t} = \kappa\left(\frac{\partial^2 u}{\partial r^2} + \frac{1}{r}\frac{\partial u}{\partial r} + \frac{1}{r^2}\frac{\partial^2 u}{\partial \theta^2}\right); \qquad . \quad (2.35)$$

and if the flow is radial, so that u is independent of θ and $\partial u/\partial\theta = 0$, it reads

$$\frac{\partial u}{\partial t} = \kappa\left(\frac{\partial^2 u}{\partial r^2} + \frac{1}{r}\frac{\partial u}{\partial r}\right). \qquad . \quad . \quad (2.36)$$

II. If the flow is *one-dimensional* and so takes place in one direction, which we take to be that of the x-axis, then $\partial u/\partial y = 0$ and $\partial u/\partial z = 0$, and the equation further reduces to

$$\frac{\partial u}{\partial t} = \kappa\frac{\partial^2 u}{\partial x^2} \qquad . \quad . \quad . \quad (2.37)$$

III. *Steady flow.*—The flow is said to be *steady* when the temperature at every point is constant, so that u is independent of the time t and is a function of the space co-ordinates only. The equation satisfied by u is then found by putting $\partial u/\partial t = 0$ in the appropriate equation above.

§ 33. *Boundary conditions.*

Besides satisfying the differential equation in the interior of the body, the temperature u must generally satisfy certain equations over the surface, usually called the *boundary conditions.* Three cases will be mentioned here :—

I. The surface may be maintained at a constant temperature, say u_0, perhaps by means of liquid at temperature u_0 flowing round it. In this case the function obtained for u in the interior of the body, a function of x, y, z and t, must reduce to u_0 when the space co-ordinates refer to a point on the surface.

More generally, the temperature over the surface may be any given function of position and time.

II. The surface of the body may be impervious to heat, in which case the condition to be satisfied over the surface will be $\partial u/\partial n = 0$, i.e. the temperature gradient will be zero in a direction normal to the surface.

III. The body may be surrounded by a gas into which heat is radiating from the surface. In this case, if u is the surface temperature of the body, and u_0 that of the surrounding gas, we make the assumption that the body loses heat at a rate proportional to $u - u_0$ and we put the rate equal to $\mathrm{H}(u - u_0)$ per unit area per unit time, where H is a constant, called the coefficient of *emissivity* or *exterior conductivity*.

With this assumption, consider the quantity of heat $d\mathrm{Q}$ gained in time dt by a coin-shaped element, of thickness ϵ, with one of its flat faces of area $d\mathrm{S}$ in the surface (Fig. 11).

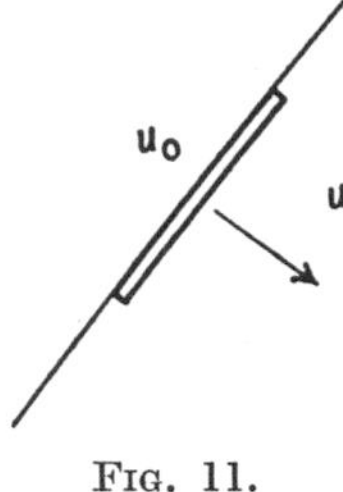

FIG. 11.

The quantity of heat that flows across the interior flat face by conduction is $\mathrm{K} \,.\, \partial u/\partial n \,.\, d\mathrm{S}\, dt$, by (2.28), where $\partial u/\partial n$ is the temperature gradient in the direction of the *inward* normal; while the quantity lost by radiation from the other flat face is $\mathrm{H}(u - u_0)d\mathrm{S}\, dt$; consequently, ignoring the flow across the narrow cylindrical surface of the element, we have

$$d\mathrm{Q} = \mathrm{K}\frac{\partial u}{\partial n}d\mathrm{S}\, dt - \mathrm{H}(u - u_0)d\mathrm{S}\, dt.$$

But $d\mathrm{Q}$ is also given, as in (2.31), by

$$d\mathrm{Q} = s \,.\, \rho\epsilon d\mathrm{S} \,.\, \frac{\partial u}{\partial t}dt.$$

The first of these expressions for $d\mathrm{Q}$ is proportional to the area of the element, the second to its volume; on equating them, dividing by $d\mathrm{S}\, dt$, and making ϵ tend to zero, we get

$$\mathrm{K}\frac{\partial u}{\partial n} = \mathrm{H}(u - u_0),$$

or

$$\frac{\partial u}{\partial n} = h(u - u_0), \qquad . \qquad . \qquad . \quad (2.38)$$

where $h = \mathrm{H}/\mathrm{K}$. This is the condition that must be satisfied when the space co-ordinates in u refer to a point on the surface.

§ 34. *Cooling of long circular cylinder.*

To take an example, consider the cooling of a long circular cylinder, initially heated to a uniform temperature u_1, when its surface is maintained at a constant temperature u_0. Let the radius of the cylinder be a, and suppose it so long that its length may be theoretically regarded as infinite. The problem is then a two-dimensional one, in which the flow of heat is radial, and the equation to be satisfied by the temperature u in the interior of the cylinder is (2.36), viz.

$$\frac{\partial u}{\partial t} = \kappa\left(\frac{\partial^2 u}{\partial r^2} + \frac{1}{r}\frac{\partial u}{\partial r}\right).$$

The temperature at the surface is supposed to adjust itself instantaneously to the value u_0, so that the boundary condition to be satisfied is

$$u = u_0, \quad r = a, \quad 0 < t < \infty.$$

The initial condition is

$$u = u_1, \quad 0 < r < a, \quad t = 0.$$

Further, u will approximate to u_0 as t increases, and will always be finite throughout the cylinder and in particular at $r = 0$.

The problem is a little simplified by first making the substitution

$$v = u - u_0 \quad . \quad . \quad . \quad (2.39)$$

Then v must satisfy the following equation and conditions :

$$\frac{\partial v}{\partial t} = \kappa\left(\frac{\partial^2 v}{\partial r^2} + \frac{1}{r}\frac{\partial v}{\partial r}\right); \quad . \quad . \quad . \quad (2.40)$$

$$v = 0, \quad r = a, \quad 0 < t < \infty ; \quad . \quad . \quad \text{(i)}$$

$$v = u_1 - u_0, \quad 0 < r < a, \quad t = 0 ; \quad . \quad \text{(ii)}$$

$$v \to 0 \text{ when } t \to \infty ; \quad . \quad . \quad . \quad . \quad \text{(iii)}$$

$$v \text{ is finite, } 0 < r < a, \ 0 < t < \infty. \quad . \quad \text{(iv)}$$

Using a standard method of seeking particular solutions of partial differential equations, we make the substitution

$$v = \mathrm{RT} \quad . \quad . \quad . \quad (2.41)$$

in the equation, where R denotes a function of r only, and T a function of t only ; the result can be written

$$\frac{1}{\kappa\mathrm{T}}\frac{d\mathrm{T}}{dt} = \frac{1}{\mathrm{R}}\left(\frac{d^2\mathrm{R}}{dr^2} + \frac{1}{r}\frac{d\mathrm{R}}{dr}\right). \quad . \quad . \quad (2.42)$$

On the face of it, one side of this equation is a function of r only, the other a function of t only, and since r and t are independent variables, an equation of such a type is impossible, unless each side is equal to the same constant. Accordingly, put each side equal to a constant λ ; then T and R must satisfy the separate equations

$$\frac{d\mathrm{T}}{dt} = \lambda\kappa\mathrm{T}, \quad . \quad . \quad . \quad (2.43)$$

$$\frac{d^2\mathrm{R}}{dr^2} + \frac{1}{r}\frac{d\mathrm{R}}{dr} = \lambda\mathrm{R}. \quad . \quad . \quad (2.44)$$

The solution of (2.43) is of the form $\mathrm{T} = \mathrm{A}e^{\lambda\kappa t}$, where A is an arbitrary constant. This form, however, is only possible if λ is negative, on account of condition (iii) ; accordingly, put $\lambda = -\mu^2$, then

$$\mathrm{T} = \mathrm{A}e^{-\mu^2\kappa t},$$

and (2.44) becomes

$$\frac{d^2\mathrm{R}}{dr^2} + \frac{1}{r}\frac{d\mathrm{R}}{dr} + \mu^2\mathrm{R} = 0,$$

from which follows, by § 7,

$$\mathrm{R} = \mathrm{BJ}_0(\mu r) + \mathrm{CY}_0(\mu r),$$

where B, C are constants, and $\mu > 0$ for $\mathrm{Y}_0(\mu r)$ to be real when $r > 0$. But since $\mathrm{Y}_0(\mu r) \to -\infty$ when $r \to 0$, we must put $\mathrm{C} = 0$ by condition (iv), and hence, merging A and B into one constant,

$$v = \mathrm{A}e^{-\mu^2\kappa t}\mathrm{J}_0(\mu r).$$

Condition (i) will also be satisfied if

$$J_0(\mu a) = 0,$$

i.e. if $\mu a = \alpha$, or $\mu = \alpha/a$, where α is a typical positive root of the equation $J_0(x) = 0$. Substituting this value of μ, we now have the solution

$$v = Ae^{-\alpha^2\kappa t/a^2} J_0\left(\frac{\alpha r}{a}\right),$$

which satisfies all the conditions except (ii). But, since (2.40) is a linear equation, the sum of any number of solutions is also a solution. Consequently we can write down the more general solution

$$v = \sum_{\alpha} Ae^{-\alpha^2\kappa t/a^2} J_0\left(\frac{\alpha r}{a}\right), \quad . \quad . \quad (2.45)$$

where the summation extends over the positive roots of the equation $J_0(x) = 0$. This solution satisfies every condition but (ii), since this is true of every term. But it can be made to satisfy (ii) also ; for, putting $t = 0$, we have only to choose the constants A so that

$$u_1 - u_0 = \sum_{\alpha} AJ_0\left(\frac{\alpha r}{a}\right), \quad (0 < r < a),$$

or, if $x = r/a$,

$$u_1 - u_0 = \sum_{\alpha} AJ_0(\alpha x), \quad (0 < x < 1).$$

By (1.48) and Exs. I, 2, (i), we find

$$A = \frac{2(u_1 - u_0)}{\alpha J_1(\alpha)},$$

and substituting this value of A in (2·45) and the resulting value of v in (2.39), we obtain the final solution

$$u = u_0 + 2(u_1 - u_0)\sum_{\alpha} \frac{1}{\alpha J_1(\alpha)} e^{-\alpha^2\kappa t/a^2} J_0\left(\frac{\alpha r}{a}\right), \quad (2.46)$$

where the summation extends over the positive roots of the equation $J_0(x) = 0$.

EXAMPLES V

1. If the initial condition is $u = u_0 + \phi(r)$, $(0 < r < a)$, and the boundary condition is $u = u_0$, $(0 < t < \infty)$, show that the solution is

$$u = u_0 + \sum_\alpha A e^{-\alpha^2 \varkappa t/a^2} J_0\left(\frac{\alpha r}{a}\right),$$

where the summation extends over the positive roots of $J_0(x) = 0$, and

$$A = \frac{2}{J_1^2(\alpha)} \int_0^1 x\phi(ax) J_0(\alpha x)\, dx.$$

2. If the initial condition is $u = u_0 + \phi(r)$, $(0 < r < a)$, and the surface is impervious to heat, so that the boundary condition is $\partial u/\partial r = 0$, $(r = a,\ 0 < t < \infty)$, show that the solution is

$$u = u_0 + A_0 + \sum_\alpha A e^{-\alpha^2 \varkappa t/a^2} J_0\left(\frac{\alpha r}{a}\right),$$

where the summation extends over the positive roots of $J_1(x) = 0$, and

$$A_0 = 2\int_0^1 x\phi(ax)\, dx, \quad A = \frac{2}{J_0^2(\alpha)} \int_0^1 x\phi(ax) J_0(\alpha x)\, dx.$$

3. If the cylinder is at a uniform temperature u_1 initially, and at time $t = 0$ is placed in a gas at temperature u_0, show that the solution is

$$u = u_0 + 2(u_1 - u_0) \sum_\alpha \frac{J_1(\alpha)}{\alpha\{J_0^2(\alpha) + J_1^2(\alpha)\}} e^{-\alpha^2 \varkappa t/a^2} J_0\left(\frac{\alpha r}{a}\right),$$

where the summation extends over the positive roots of the equation

$$x J_0'(x) + ah J_0(x) = 0,$$

and h has the same meaning as in (2.38).

CHAPTER III

MODIFIED BESSEL FUNCTIONS

§ 35. *Modified Bessel functions of zero order.*

If we put $k = i = \sqrt{(-1)}$ in (1.18), we obtain the equation

$$\frac{d^2y}{dx^2} + \frac{1}{x}\frac{dy}{dx} - y = 0, \quad . \quad . \quad . \quad (3.1)$$

which is called the *modified Bessel equation* of zero order; it can also be written

$$\frac{d}{dx}\left(x\frac{dy}{dx}\right) = xy. \quad . \quad . \quad . \quad (3.2)$$

A solution of this equation, and the only one, except for a constant factor, that remains finite when $x = 0$, is denoted by $I_0(x)$ and is given by *

$$I_0(x) = J_0(ix) = 1 + \frac{x^2}{2^2} + \frac{x^4}{2^2 . 4^2} + \frac{x^6}{2^2 . 4^2 . 6^2} + \ldots \quad (3.3)$$

which is called the *modified Bessel function of the first kind* of zero order.

§ 36. A modified Bessel function *of the second kind* can now be defined as any solution of (3.1) which is not a constant multiple of $I_0(x)$, and can be expressed in the form

$$AI_0(x) + BI_0(x)\int\frac{dx}{xI_0{}^2(x)},$$

(cf. § 4), where A, B are any constants ($B \neq 0$).

When x is small this solution (cf. § 5) behaves like

$$AI_0(x) + B\left\{I_0(x)\log x - \frac{x^2}{4} - \ldots\right\}$$

* Cf. $\cosh x = \cos ix$.

In particular, if we put

$$A = \log 2 - \gamma, \quad B = -1,$$

where γ denotes Euler's constant (§ 6), we obtain a particular modified Bessel function of the second kind which is denoted by $K_0(x)$, thus

$$K_0(x) = (\log 2 - \gamma)I_0(x) - \left\{I_0(x) \log x - \frac{x^2}{4} - \ldots\right\}. \quad (3.4)$$

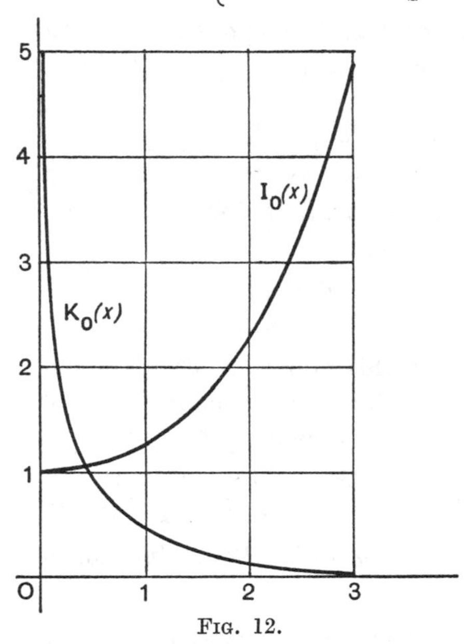

FIG. 12.

We note that, when x is small,

$$K_0(x) = (\log 2 - \gamma) - \log x \ldots \quad . \quad (3.5)$$

the remaining terms being small in comparison with unity; so that $K_0(x) \to +\infty$ when $x \to +0$.

The graphs of $I_0(x)$, $K_0(x)$ are shown together in Fig. 12.

Ex. 1. Show that the equation $I_0(x) = 0$ has no real root; and that the equation $I_0'(x) = 0$ has no real root except $x = 0$.

Ex. 2. Show that

$$K_0(x) = \frac{\pi i}{2}\{J_0(ix) + iY_0(ix)\},$$

provided that log (ix), which occurs in $Y_0(ix)$, has the value $\frac{1}{2}\pi i + \log x$ when x is positive.

§ 37. The general solution of (3.1) can now be written

$$y = AI_0(x) + BK_0(x), \qquad (3.6)$$

where A, B are arbitrary constants.

Corollary. The general solution of the equation

$$\frac{d^2y}{dx^2} + \frac{1}{x}\frac{dy}{dx} - k^2y = 0,$$

where k is a constant, can be written

$$y = AI_0(kx) + BK_0(kx), \qquad (3.7)$$

where $k > 0$ for $K_0(kx)$ to be real when $x > 0$.

§ 38. Since $I_0(x)$ is a solution of (3.2), we have

$$\frac{d}{dx}\left(x\frac{dI_0(x)}{dx}\right) = xI_0(x), \qquad (3.8)$$

and inversely

$$\int xI_0(x)dx = xI_0'(x). \qquad (3.9)$$

Replacing x by αx, we have further

$$\int xI_0(\alpha x)dx = \frac{x}{\alpha}I_0'(\alpha x), \qquad (3.10)$$

and, as in § 11, we find, if α, β are constants,

$$(\beta^2 - \alpha^2)\int xI_0(\alpha x)I_0(\beta x)dx$$
$$= x\{\beta I_0'(\beta x)I_0(\alpha x) - \alpha I_0'(\alpha x)I_0(\beta x)\}, \quad (3.11)$$

$$\int xI_0^2(\alpha x)dx = \frac{x^2}{2}\{I_0^2(\alpha x) - I_0'^2(\alpha x)\}. \qquad (3.12)$$

§ 39. *Laplace's equation in cylindrical co-ordinates, when the dependent variable is independent of θ.*

When Laplace's equation

$$\frac{\partial^2 u}{\partial x^2} + \frac{\partial^2 u}{\partial y^2} + \frac{\partial^2 u}{\partial z^2} = 0 \qquad (3.13)$$

is expressed in terms of cylindrical co-ordinates r, θ, z, it reads

$$\frac{\partial^2 u}{\partial r^2} + \frac{1}{r}\frac{\partial u}{\partial r} + \frac{1}{r^2}\frac{\partial^2 u}{\partial \theta^2} + \frac{\partial^2 u}{\partial z^2} = 0, \qquad (3.14)$$

and if u is independent of θ, reduces to

$$\frac{\partial^2 u}{\partial r^2} + \frac{1}{r}\frac{\partial u}{\partial r} + \frac{\partial^2 u}{\partial z^2} = 0. \quad . \quad . \quad (3.15)$$

We shall now find solutions of this equation of the form

$$u = \mathrm{RZ},$$

where R is a function of r only, and Z a function of z only. When we substitute $u = \mathrm{RZ}$ in (3.15), we find that the resulting equation can be written

$$\frac{1}{\mathrm{R}}\left(\frac{d^2\mathrm{R}}{dr^2} + \frac{1}{r}\frac{d\mathrm{R}}{dr}\right) = -\frac{1}{\mathrm{Z}}\frac{d^2\mathrm{Z}}{dz^2},$$

each side of which must be equal to the same constant, for the same kind of reason as in § 34. Firstly, putting each side equal to $-\mu^2$, we find solutions of the form

$$u = \{\mathrm{AJ}_0(\mu r) + \mathrm{BY}_0(\mu r)\}(\mathrm{C} \sinh \mu z + \mathrm{D} \cosh \mu z), \quad (3.16)$$

or the equivalent form

$$u = \{\mathrm{AJ}_0(\mu r) + \mathrm{BY}_0(\mu r)\}(\mathrm{C}e^{-\mu z} + \mathrm{D}e^{\mu z}). \quad (3.17)$$

Secondly, putting each side equal to $+\mu^2$, we find

$$u = \{\mathrm{AI}_0(\mu r) + \mathrm{BK}_0(\mu r)\}(\mathrm{C} \sin \mu z + \mathrm{D} \cos \mu z). \quad (3.18)$$

Thirdly, putting each side equal to zero, we find

$$u = (\mathrm{A} + \mathrm{B} \log r)(\mathrm{C}z + \mathrm{D}). \quad . \quad . \quad (3.19)$$

In each case, A, B, C, D denote arbitrary constants. We must suppose $\mu > 0$ for $\mathrm{Y}_0(\mu r)$ or $\mathrm{K}_0(\mu r)$, to be real when $r > 0$.

§ 40. *Steady flow of heat in a finite cylinder.*

Consider the flow of heat in a finite cylinder of radius a and length l. Suppose that over one end the temperature is a given function of the distance from the centre of that end, and that the other end and the curved surface are maintained at a constant temperature, which we take as the zero temperature. Further, suppose that these conditions have persisted for some time, so that a steady state has been reached in which $\partial u/\partial t = 0$ at every point (§ 32, III).

Then, putting $\partial u/\partial t = 0$ in (2.32), we see that u satisfies Laplace's equation. Cylindrical co-ordinates are appropriate to the present problem, with the pole at the centre of one of the ends and the z-axis along the axis of the cylinder. Then, as u does not depend on θ, the equation satisfied by u is (3.15), viz.

$$\frac{\partial^2 u}{\partial r^2} + \frac{1}{r}\frac{\partial u}{\partial r} + \frac{\partial^2 u}{\partial z^2} = 0. \qquad (3.20)$$

The boundary conditions we take to be

$$u = 0, \quad (z = 0, \quad 0 < r < a), \qquad \text{(i)}$$
$$u = 0, \quad (r = a, \quad 0 < z < l), \qquad \text{(ii)}$$
$$u = \phi(r), \quad (z = l, \quad 0 < r < a), \qquad \text{(iii)}$$
$$u \text{ is finite}, \quad (0 < r < a, \quad 0 < z < l). \qquad \text{(iv)}$$

With these conditions in mind, we select from (3.16), (3.18), (3.19) the types of solution likely to be suitable. Firstly, having regard to (iv), we must put $B = 0$ in each case, since $Y_0(\mu r)$, $K_0(\mu r)$, and $\log r$ all become infinite at $r = 0$.

Next, having regard to (i), we put $D = 0$ in each case. We then have possible solutions of the form

$$J_0(\mu r) \sinh \mu z, \quad I_0(\mu r) \sin \mu z, \quad z.$$

Of these three solutions, the first will satisfy (ii) if

$$J_0(\mu a) = 0,$$

that is, if $\mu a = \alpha$, or $\mu = \alpha/a$, where α is any positive root of the equation $J_0(x) = 0$. The second cannot satisfy (ii) because the equation $I_0(x) = 0$ has no real root (§ 36, Ex. 1); and the third obviously cannot satisfy (ii). Hence we select the solution

$$u = AJ_0\left(\frac{\alpha r}{a}\right) \sinh \frac{\alpha z}{a}$$

which satisfies all the conditions but (iii). The same is true of the more general solution

$$u = \sum_{\alpha} AJ_0\left(\frac{\alpha r}{a}\right) \sinh \frac{\alpha z}{a}, \qquad (3.21)$$

where the summation extends over all the positive roots of $J_0(x) = 0$. Further, this more general solution will also satisfy (iii) provided the constants A are determined so that (putting $z = l$)

$$\phi(r) = \sum_\alpha AJ_0\left(\frac{\alpha r}{a}\right) \sinh \frac{\alpha l}{a}, \quad (0 < r < a),$$

or, if $r = ax$,

$$\phi(ax) = \sum_\alpha AJ_0(\alpha x) \sinh \frac{\alpha l}{a}, \quad (0 < x < 1).$$

Hence, by (1.48),

$$A \sinh\frac{\alpha l}{a} = \frac{2}{J_1^2(\alpha)} \int_0^1 x\phi(ax)J_0(\alpha x)dx \quad . \quad (3.22)$$

When the value of A, thus found, is substituted in (3.21), we obtain the final solution.

Examples VI

1. If $\phi(r) = u_0$, a constant, show that the solution is

$$u = 2u_0 \sum_\alpha \frac{J_0\left(\frac{\alpha r}{a}\right) \sinh \frac{\alpha z}{a}}{\alpha J_1(\alpha) \sinh \frac{\alpha l}{a}},$$

where $J_0(\alpha) = 0$.

2. If the boundary conditions are $u = u_0$, $(z = 0,\ 0 < r < a)$; $u = 0$, $(r = a,\ 0 < z < l)$; $u = 0$, $(z = l,\ 0 < r < a)$; show that the solution is

$$u = 2u_0 \sum_\alpha \frac{J_0\left(\frac{\alpha r}{a}\right) \sinh \frac{\alpha(l - z)}{a}}{\alpha J_1(\alpha) \sinh \frac{\alpha l}{a}},$$

where $J_0(\alpha) = 0$.

3. If the boundary conditions are $u = 0$, $(z = 0,\ 0 < r < a)$; $u = 0$, $(z = l,\ 0 < r < a)$; $u = u_0$, $(r = a,\ 0 < z < l)$; show that the solution is

$$u = \frac{4u_0}{\pi} \sum_{s=1}^{\infty} \frac{I_0\left(\frac{m\pi r}{l}\right) \sin \frac{m\pi z}{l}}{mI_0\left(\frac{m\pi a}{l}\right)},$$

where $m = 2s - 1$.

4. Show how to solve the problem when the boundary conditions are $u = \psi(r)$, $(z = 0,\ 0 < r < a)$; $u = \phi(r)$, $(z = l,\ 0 < r < a)$; $u = f(z)$, $(r = a,\ 0 < z < l)$.

5. If the boundary conditions are $u = 0$, $(z = 0,\ 0 < r < a)$; $u = \phi(r)$, $(z = l,\ 0 < r < a)$; $\partial u/\partial r = 0$, $(r = a,\ 0 < z < l)$, so that the curved surface is impervious to heat, show that the solution is

$$u = A_0 z + \sum_{\alpha} A J_0\left(\frac{\alpha r}{a}\right) \sinh \frac{\alpha z}{a},$$

where the summation extends over the positive roots of the equation $J_1(x) = 0$, and the coefficients A_0, A are given by

$$\tfrac{1}{2} l A_0 = \int_0^1 x\phi(ax)dx,$$

$$\tfrac{1}{2} A J_0^2(\alpha) \sinh \frac{\alpha l}{a} = \int_0^1 x\phi(ax) J_0(\alpha x)dx.$$

Verify that, if $\phi(r) = u_0$, a constant, this solution reduces to $u = u_0 z/l$, as is obvious from physical considerations.

§ 41. *Large values of x.*

It is plain from the series (3.3) that $I_0(x)$ tends to $+\infty$ when x is large. In what follows we shall need an approximation to $I_0(x)$ when x is large and positive. To obtain such an approximation, we first put

$$u = y\sqrt{x} \qquad . \quad . \quad . \quad (3.23)$$

in (3.1), and find that u satisfies the equation

$$\frac{d^2u}{dx^2} = \left(1 - \frac{1}{4x^2}\right)u \qquad . \quad . \quad . \quad (3.24)$$

which, when x is large compared with 1, takes the approximate form $u'' = u$, of which the general solution is $u = Ae^x + Be^{-x}$, where A, B are arbitrary constants. This suggests that any solution of (3.24), which tends to ∞ when $x \to +\infty$, will approximate to Ae^x when x is large and positive.

§ 42. To attempt to improve upon this approximation, we make the substitution

$$u = ve^x$$

in (3.24), and find that v satisfies the equation

$$\frac{d^2v}{dx^2} + 2\frac{dv}{dx} + \frac{v}{4x^2} = 0. \qquad . \quad . \quad (3.25)$$

Assuming that this equation can be satisfied by a series of the form

$$v = 1 + \frac{c_1}{x} + \frac{c_2}{x^2} + \frac{c_3}{x^3} + \ldots \qquad . \quad (3.26)$$

we substitute this series for v in the equation and obtain, after collecting like terms,

$$\left\{2c_1 - \left(\frac{1}{2}\right)^2\right\}\frac{1}{x^2} + \left\{2 \,.\, 2c_2 - \left(\frac{3}{2}\right)^2 c_1\right\}\frac{1}{x^3}$$
$$+ \left\{2 \,.\, 3c_3 - \left(\frac{5}{2}\right)^2 c_2\right\}\frac{1}{x^4} + \ldots = 0.$$

By equating the several coefficients to zero, it follows that the equation is formally satisfied by the series, provided the coefficients are given by

$$c_1 = \frac{1}{8},$$
$$c_2 = \frac{3^2}{2 \,.\, 8}c_1 = \frac{1^2 \,.\, 3^2}{2 \,!\, 8^2},$$
$$c_3 = \frac{5^2}{3 \,.\, 8}c_2 = \frac{1^2 \,.\, 3^2 \,.\, 5^2}{3 \,!\, 8^3},$$
$$\ldots\ldots$$

and hence

$$v = 1 + \frac{1^2}{8x} + \frac{1^2 \,.\, 3^2}{2 \,!\, (8x)^2} + \frac{1^2 \,.\, 3^2 \,.\, 5^2}{3 \,!\, (8x)^3} + \ldots \qquad (3.27)$$

We are thus led to the expansion

$$I_0(x) = \frac{Ae^x}{\sqrt{x}}\left(1 + \frac{1^2}{8x} + \frac{1^2 \,.\, 3^2}{2 \,!\, (8x)^2} + \frac{1^2 \,.\, 3^2 \,.\, 5^2}{3 \,!\, (8x)^3} + \ldots\right), \quad (3.28)$$

where A is some positive constant; it will be seen later (§ 81) that $A = 1/\sqrt{(2\pi)}$.

When A has this value, (3.28) is an *asymptotic expansion* (§ 78) of $I_0(x)$; the series on the right is divergent, but it has the property that the sum of the first n terms gives an approximation to $I_0(x)$ when x is large enough, with a percentage error as small as we please.

Ex. 1. Show that, when x is small,

$$\frac{xI_0(x)}{2I_0'(x)} = 1 + \frac{x^2}{8} - \frac{1}{3}\left(\frac{x^2}{8}\right)^2 + \frac{1}{6}\left(\frac{x^2}{8}\right)^3 \ldots$$

Ex. 2. Show that, when x is large,

$$\frac{I_0(x)}{I_0'(x)} = 1 + \frac{1}{2x} + \frac{3}{8x^2} \ldots$$

§ 43. *Application to alternating current in a wire of circular cross section.*

The differential equations of the electromagnetic field are based upon two laws which are sometimes distinguished by the names of Ampère and Faraday, viz.

Ampère's law.—The line integral of magnetic force round a closed circuit is equal to $4\pi \times$ (the integral of electric current through the circuit).

Faraday's law.—The line integral of electric force round a closed circuit is equal to $-\frac{\partial}{\partial t}$ (magnetic induction through the circuit).

§ 44. To apply these laws to determine the current density at radius r in a wire of circular cross-section, through which alternating current is flowing, let a be the radius of the wire, ρ its specific resistance, and μ its permeability; let x be the current density and H the magnetic intensity at radius r and time t.

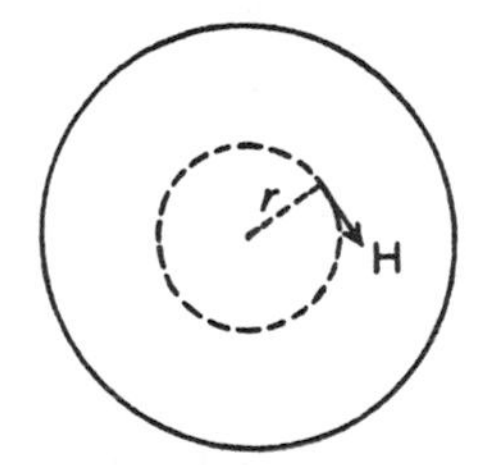

Fig. 13.

Firstly, consider a closed circuit which is a circle of radius r, with its axis along the axis of the wire (Fig. 13). Applying Ampère's law to this circuit, we have

$$2\pi r\mathrm{H} = 4\pi\int_0^r x \,.\, 2\pi r dr, \quad . \quad . \quad (3.29)$$

and hence, after differentiating with regard to r,

$$\frac{1}{r}\frac{\partial}{\partial r}(r\mathrm{H}) = 4\pi x. \quad . \quad . \quad . \quad (3.30)$$

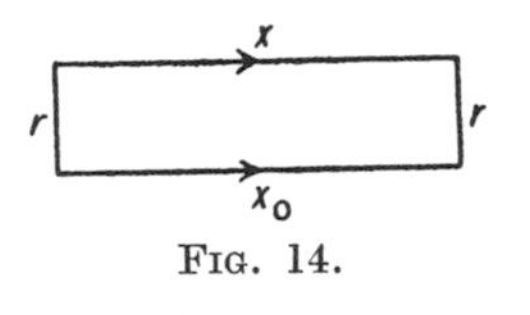

FIG. 14.

Secondly, consider a closed circuit which is a rectangle with one of its sides, of length l, along the axis of the cylinder and the two perpendicular sides of length r (Fig. 14). Applying Faraday's law, we have

$$\rho x_0 l - \rho x l = -\frac{\partial}{\partial t}\int_0^r \mu \mathrm{H} l \, dr,$$

where x_0 denotes the value of x when $r = 0$; and by differentiation with regard to r,

$$\rho\frac{\partial x}{\partial r} = \mu\frac{\partial \mathrm{H}}{\partial t}. \quad . \quad . \quad . \quad (3.31)$$

To eliminate H, multiply (3.31) by r, differentiate with regard to r, and use (3.30); this leads to the equation

$$\frac{1}{r}\frac{\partial}{\partial r}\left(r\frac{\partial x}{\partial r}\right) = \frac{4\pi\mu}{\rho}\frac{\partial x}{\partial t}, \quad . \quad . \quad (3.32)$$

from which x can be found, and then H is given by (3.29).

§ 45. Let the total current through the wire be C cos ωt, an alternating current of period $2\pi/\omega$. It is convenient to regard this current as the real part of the complex number $\mathrm{C}e^{i\omega t}$, and correspondingly to regard x as the real part of a complex number z that satisfies the equation

$$\frac{1}{r}\frac{\partial}{\partial r}\left(r\frac{\partial z}{\partial r}\right) = \frac{4\pi\mu}{\rho}\frac{\partial z}{\partial t}. \quad . \quad . \quad (3.33)$$

Accordingly, we seek a solution of this equation of the form

$$z = \mathrm{F}(r)e^{i\omega t}, \quad . \quad . \quad . \quad (3.34)$$

remembering that when F(r) has been found, the real part of z will be the actual current density.

Substituting (3.34) in (3.33), we find that $F(r)$ must satisfy the equation

$$\frac{1}{r}\frac{d}{dr}\left(r\frac{dF}{dr}\right) = \frac{4\pi\mu\omega i}{\rho}F, \quad . \quad . \quad (3.35)$$

or

$$\frac{d^2F}{dr^2} + \frac{1}{r}\frac{dF}{dr} - k^2F = 0,$$

where F denotes $F(r)$, and

$$k^2 = \frac{4\pi\mu\omega i}{\rho}, \quad k = \frac{1+i}{\sqrt{2}}\left(\frac{4\pi\mu\omega}{\rho}\right)^{\frac{1}{2}}. \quad . \quad (3.36)$$

Hence, by § 37,

$$F(r) = AI_0(kr) + BK_0(kr).$$

The constant B must be zero, since z is finite at $r = 0$, and therefore

$$z = AI_0(kr)e^{i\omega t}. \quad . \quad . \quad . \quad (3.37)$$

The constant A can be found in terms of C. For, since $Ce^{i\omega t}$ is the total current, we have

$$Ce^{i\omega t} = \int_0^a z \,.\, 2\pi r\, dr$$

and therefore

$$C = 2\pi A\int_0^a rI_0(kr)dr = 2\pi A\frac{aI_0'(ka)}{k},$$

by (3.10). This gives A in terms of C, and hence

$$z = \frac{kC}{2\pi aI_0'(ka)}I_0(kr)e^{i\omega t}. \quad . \quad . \quad (3.38)$$

Ex. Show that H is the real part of

$$\frac{2C}{aI_0'(ka)}I_0'(kr)e^{i\omega t}.$$

§ 46. *Equivalent resistance and internal self-inductance of a length l of the wire.*

The electromotive force along a length l of the surface of the wire, where $r = a$, is the real part of

$$\rho l \,.\, \frac{kCI_0(ka)}{2\pi aI_0'(ka)}e^{i\omega t}.$$

If we equate this to $(\mathrm{R} + i\omega\mathrm{L})\mathrm{C}e^{i\omega t}$, then R is called the *equivalent resistance* and L the *internal self-inductance* of length l of the wire ; this gives

$$\mathrm{R} + i\omega\mathrm{L} = \frac{\rho l k \mathrm{I}_0(ka)}{2\pi a \mathrm{I}_0'(ka)}.$$

Further, if R_0 is the resistance of length l for steady current,

$$\mathrm{R}_0 = \frac{\rho l}{\pi a^2}$$

and, by division,

$$\frac{\mathrm{R} + i\omega\mathrm{L}}{\mathrm{R}_0} = \frac{ka}{2}\frac{\mathrm{I}_0(ka)}{\mathrm{I}_0'(ka)}. \quad . \quad . \quad (3.39)$$

§ 47. *Low frequency.*

Now, if we put (3.36) in the form

$$\frac{k^2a^2}{8} = \frac{\pi\mu\omega a^2}{2\rho}i = i\kappa^2, \quad \kappa = \left(\frac{\pi\mu\omega a^2}{2\rho}\right)^{\frac{1}{2}},$$

we find, from (3.39), using § 42, Ex. 1, when the frequency is small (ω small, κ small),

$$\frac{\mathrm{R} + i\omega\mathrm{L}}{\mathrm{R}_0} = 1 + i\kappa^2 + \frac{\kappa^4}{3} - \frac{i\kappa^6}{6} \cdots$$

and hence, by equating real and imaginary parts,

$$\frac{\mathrm{R}}{\mathrm{R}_0} = 1 + \frac{\kappa^4}{3} \cdots, \quad \frac{\omega\mathrm{L}}{\mathrm{R}_0} = \kappa^2 - \frac{\kappa^6}{6} \cdots$$

or, since $\mathrm{R}_0 = \rho l/\pi a^2$, and $\omega = 2\rho\kappa^2/\pi\mu a^2$,

$$\mathrm{R} = \mathrm{R}_0\left(1 + \frac{\kappa^4}{3} \cdots\right), \quad \mathrm{L} = \frac{l\mu}{2}\left(1 - \frac{\kappa^4}{6} \cdots\right). \quad (3.40)$$

§ 48. *High frequency.*

Again, if the frequency is large (ω large, κ large) we have, from (3.39), using § 42, Ex. 2,

$$\frac{\mathrm{R} + i\omega\mathrm{L}}{\mathrm{R}_0} = \frac{ka}{2} + \frac{1}{4} + \frac{3}{16ka} \cdots$$

$$= (1 + i)\kappa + \frac{1}{4} + \frac{3}{32(1 + i)\kappa} \cdots$$

and hence, by equating real and imaginary parts,

$$\frac{\mathrm{R}}{\mathrm{R}_0} = \kappa + \frac{1}{4} + \frac{3}{64\kappa}\ \ldots, \quad \frac{\omega \mathrm{L}}{\mathrm{R}_0} = \kappa - \frac{3}{64\kappa}\ \ldots$$

or

$$\mathrm{R} = \mathrm{R}_0\left(\kappa + \frac{1}{4} + \frac{3}{64\kappa}\ \ldots\right), \quad \mathrm{L} = \frac{l\mu}{2}\left(\frac{1}{\kappa} - \frac{3}{64\kappa^3}\ \ldots\right) \tag{3.41}$$

§ 49. *Verification of the value found for the equivalent high-frequency resistance.*

As a further example, we may verify the formula just obtained for the resistance R, when the frequency is high, by showing that the heat Q, generated per unit time in length l, is $\frac{1}{2}\mathrm{RC}^2$. We have, in fact, x being the real current density,

$$\mathrm{Q} = \rho l \int_0^1 \int_0^a x^2 \,.\, 2\pi r\, dr\, dt = 2\pi\rho l \int_0^1 \int_0^a \left(\frac{z + \bar{z}}{2}\right)^2 r\, dr\, dt,$$

where $\bar{z}$ denotes the complex number conjugate to z. Now for rapidly alternating current, the average value of z^2 or $\bar{z}^2$ with respect to t, over an interval of a second, is practically zero, and hence

$$\mathrm{Q} = \pi\rho l \int_0^1 \int_0^a z\bar{z} r\, dr\, dt,$$

that is, substituting the value of z from (3.38), and writing $\bar{k}$ for the conjugate of k,

$$\mathrm{Q} = \frac{\pi\rho l \mathrm{C}^2}{4\pi^2 a^2} \frac{k\bar{k}}{\mathrm{I}_0'(ka)\mathrm{I}_0'(\bar{k}a)} \int_0^1 dt \int_0^a \mathrm{I}_0(kr)\mathrm{I}_0(\bar{k}r) r\, dr.$$

Using (3.11), and putting $q = ka$, $\bar{q} = \bar{k}a$, we can write the result of the integration

$$\mathrm{Q} = \frac{\rho l \mathrm{C}^2}{4\pi a^2} \frac{(q\bar{q})^2}{q^2 - \bar{q}^2}\left(\frac{\mathrm{I}_0(\bar{q})}{\bar{q}\mathrm{I}_0'(\bar{q})} - \frac{\mathrm{I}_0(q)}{q\mathrm{I}_0'(q)}\right),$$

or, since $q = ka = 2(1 + i)\kappa$,

$$\mathrm{Q} = \frac{\rho l \mathrm{C}^2 \kappa^2 i}{\pi a^2}\left(\frac{\mathrm{I}_0(q)}{q\mathrm{I}_0'(q)} - \frac{\mathrm{I}_0(\bar{q})}{\bar{q}\mathrm{I}_0'(\bar{q})}\right),$$

and therefore, by § 42, Ex. 2, since κ is large for rapidly alternating current,

$$\begin{aligned} Q &= \frac{\rho l C^2 \kappa^2 i}{\pi a^2}\left(\frac{1}{q} + \frac{1}{2q^2} + \frac{3}{8q^2} \cdots - \frac{1}{\bar{q}} - \frac{1}{2\bar{q}^2} - \frac{3}{8\bar{q}^2} \cdots\right) \\ &= \frac{\rho l C^2}{\pi a^2}\left(\frac{\kappa}{2} + \frac{1}{8} + \frac{3}{128\kappa} \cdots\right) \\ &= \frac{1}{2}R_0 C^2\left(\kappa + \frac{1}{4} + \frac{3}{64\kappa} \cdots\right) = \frac{1}{2}RC^2, \end{aligned}$$

which is the result we set out to verify.

§ 50. *The skin effect.*

To conclude this application to the flow of alternating current in a cylindrical wire, we shall verify that, for sufficiently high frequencies, the current flowing through a coaxial cylinder of radius r is small compared with the total current, even when r is nearly equal to a, thus showing that most of the current flows through a thin layer at the surface—the well-known " skin effect."

By (3.38) the whole current flowing through a coaxial cylinder of radius r is the real part of

$$\frac{kCe^{i\omega t}}{2\pi a I_0'(ka)}\int_0^r I_0(kr) \,.\, 2\pi r\,dr = Ce^{i\omega t}\frac{rI_0'(kr)}{aI_0'(ka)}.$$

The ratio of this to the total current $Ce^{i\omega t}$ is

$$\frac{rI_0'(kr)}{aI_0'(ka)},$$

and, by (3.28), when k is large, this is approximately equal to

$$\frac{r}{a}\frac{e^{kr}}{\sqrt{(kr)}}\frac{\sqrt{(ka)}}{e^{ka}}, \quad = \left(\frac{r}{a}\right)^{\frac{1}{2}} e^{-k(a-r)},$$

which, for any fixed value of r, however small $a - r$ may be, is as small as we please when k is sufficiently large, i.e. when the frequency is high enough.

§ 51. *Kelvin's ber and bei functions.*

The functions named ber x and bei x by Lord Kelvin may be defined by

$$\mathrm{I}_0\left(\frac{1+i}{\sqrt{2}}x\right) = \operatorname{ber} x + i \operatorname{bei} x.$$

Now, since $\{(1+i)x/\sqrt{2}\}^2 = ix^2$, we have, by (3.3),

$$\mathrm{I}_0\left(\frac{1+i}{\sqrt{2}}x\right) = 1 + \frac{ix^2}{2^2} + \frac{(ix^2)^2}{2^2 \cdot 4^2} + \frac{(ix^2)^3}{2^2 \cdot 4^2 \cdot 6^2} + \dots$$

and hence, by equating real and imaginary parts,

$$\operatorname{ber} x = 1 - \frac{x^4}{2^2 \cdot 4^2} + \frac{x^8}{2^2 \cdot 4^2 \cdot 6^2 \cdot 8^2} - \dots \qquad (3.42)$$

$$\operatorname{bei} x = \frac{x^2}{2^2} - \frac{x^6}{2^2 \cdot 4^2 \cdot 6^2} + \frac{x^{10}}{2^2 \cdot 4^2 \cdot 6^2 \cdot 8^2 \cdot 10^2} - \dots \qquad (3.43)$$

§ 52. If we now write (3.36) in the form

$$k = \frac{1+i}{\sqrt{2}}m, \quad m = \left(\frac{4\pi\mu\omega}{\rho}\right)^{\frac{1}{2}}, \qquad (3.44)$$

the results of the application begun in § 44 can be written in terms of the ber and bei functions. Thus (3.38) will be found to be expressible in the form

$$z = \frac{im\mathrm{C}}{2\pi a} \frac{\operatorname{ber} mr + i \operatorname{bei} mr}{\operatorname{ber}' ma + i \operatorname{bei}' ma} e^{i\omega t}, \qquad (3{\cdot}45)$$

and (3.39) in the form

$$\frac{\mathrm{R} + i\omega\mathrm{L}}{\mathrm{R}_0} = \frac{ima}{2} \frac{\operatorname{ber} ma + i \operatorname{bei} ma}{\operatorname{ber}' ma + i \operatorname{bei}' ma}. \qquad (3.46)$$

In these forms the real and imaginary parts are in evidence. Numerical calculation can be made with the aid of the tables published in Kelvin: "Math. and Phys. Papers," III, p. 493; Jahnke und Emde: "Funktionentafeln"; McLachlan: "Bessel Functions for Engineers," etc.

§ 53. The reader will realise that a complete familiarity with the behaviour of the function $\mathrm{J}_0(x)$ for all values of x, real or complex, would render it unnecessary to give separate names to the functions $I_0(x)$, ber x, bei x, etc.

Ex. 1. Given that $y = I_0\left(\frac{1+i}{\sqrt{2}}x\right)$ is a solution of the equation

$$\frac{1}{x}\frac{d}{dx}\left(x\frac{dy}{dx}\right) = iy,$$

by putting $y = u + iv$ in this equation, separating real and imaginary parts, and eliminating u and v in turn, show that ber x and bei x both satisfy the equation

$$\frac{1}{x}\frac{d}{dx}\left[x\frac{d}{dx}\left\{\frac{1}{x}\frac{d}{dx}\left(x\frac{dy}{dx}\right)\right\}\right] = -y.$$

Ex. 2. If α, β are constants, show that

$$\int_0^1 xJ_0(\alpha x)I_0(\beta x)\,dx = \frac{\beta I_0'(\beta)J_0(\alpha) - \alpha J_0'(\alpha)I_0(\beta)}{\alpha^2 + \beta^2};$$

and obtain the Fourier-Bessel expansion

$$I_0(kx) = 2I_0(k)\sum_{\alpha}\frac{\alpha J_0(\alpha x)}{(\alpha^2 + k^2)J_1(\alpha)},$$

where $J_0(\alpha) = 0$, $0 < x < 1$.

Deduce the Fourier-Bessel expansions of ber kx and bei kx.

CHAPTER IV

DEFINITE INTEGRALS

§ 54. *Bessel's integral for* $J_0(x)$.

If we expand the function $e^{ix\sin\theta}$ in ascending powers of x, we get

$$e^{ix\sin\theta} = 1 + \frac{ix\sin\theta}{1!} + \frac{(ix\sin\theta)^2}{2!} + \frac{(ix\sin\theta)^3}{3!} + \ldots$$

Now integrate both sides between the limits $\theta = 0$, $\theta = 2\pi$, and use the formulæ

$$\int_0^{2\pi} \sin^n\theta\, d\theta = 0, \quad (n \text{ odd}),$$

$$= \frac{(n-1)(n-3)\ \ldots\ 3\,.\,1}{n(n-2)\ \ldots\ 4\,.\,2}\,.\,2\pi, \quad (n \text{ even});$$

this gives

$$\int_0^{2\pi} e^{ix\sin\theta} d\theta = 2\pi\left(1 - \frac{x^2}{2^2} + \frac{x^4}{2^2\,.\,4^2} - \ldots\right) \quad . \quad (4.1)$$

and hence, since the series in brackets is $J_0(x)$,

$$J_0(x) = \frac{1}{2\pi}\int_0^{2\pi} e^{ix\sin\theta} d\theta. \quad . \quad . \quad (4.2)$$

Equivalent forms are given by

$$J_0(x) = \frac{1}{\pi}\int_{-\frac{\pi}{2}}^{\frac{\pi}{2}} e^{ix\sin\theta} d\theta = \frac{1}{\pi}\int_0^{\pi} e^{ix\cos\theta} d\theta, \quad . \quad (4.3)$$

or, on separating the real parts,

$$J_0(x) = \frac{2}{\pi}\int_0^{\frac{\pi}{2}} \cos(x\sin\theta) d\theta = \frac{2}{\pi}\int_0^{\frac{\pi}{2}} \cos(x\cos\theta) d\theta. \quad (4.4)$$

Any one of the above definite-integral forms of $J_0(x)$ may be called *Bessel's integral for* $J_0(x)$,* being a particular case (when $n = 0$) of Bessel's integral for $J_n(x)$, (§ 86).

§ 55. *Lipschitz's integral.*

If, in the well-known integral

$$\int_0^\infty e^{-ax} \cos bx\, dx = \frac{a}{a^2 + b^2}, \quad (a > 0),$$

we replace b by $b \cos \theta$, we get

$$\int_0^\infty e^{-ax} \cos (bx \cos \theta) dx = \frac{a}{a^2 + b^2 \cos^2 \theta}.$$

Since the infinite integral on the left is uniformly convergent with respect to θ, we may integrate under the integral sign with respect to θ from 0 to $\frac{1}{2}\pi$; we thus find

$$\int_0^\infty e^{-ax}\, dx \int_0^{\frac{\pi}{2}} \cos (bx \cos \theta)\, d\theta = \int_0^{\frac{\pi}{2}} \frac{a\, d\theta}{a^2 + b^2 \cos^2 \theta},$$

and hence by (4.4), if $a > 0$,

$$\int_0^\infty e^{-ax} J_0(bx) dx = \frac{1}{\sqrt{(a^2 + b^2)}} \quad . \quad . \quad (4.5)$$

which is known as *Lipschitz's integral.*†

COROLLARY. When $a \to 0$ we get, if $b > 0$,

$$\int_0^\infty J_0(bx) dx = \frac{1}{b} \quad . \quad . \quad . \quad (4.6)$$

and in particular, if $b = 1$,

$$\int_0^\infty J_0(x) dx = 1. \quad . \quad . \quad . \quad (4.7)$$

§ 56. *Weber's discontinuous integrals.*

Interchanging a and b in (4.5) gives, if $b > 0$,

$$\int_0^\infty J_0(ax) \,.\, e^{-bx} dx = \frac{1}{\sqrt{(a^2 + b^2)}}. \quad . \quad (4.8)$$

* Also called Parseval's integral, for historical reasons (Watson, p. 21).

† Watson, p. 384.

Since both sides of this equation are analytic functions of b when the real part of b is positive, it follows that, if $b > 0$,

$$\int_0^\infty J_0(ax) \,.\, e^{-(b+ic)x}\,dx = \frac{1}{\sqrt{\{a^2 + (b+ic)^2\}}} \qquad (4.9)$$

and hence, if $b > 0$,

$$\int_0^\infty J_0(ax) \,.\, e^{-(b+ic)x}\,dx = \frac{1}{X + iY} = \frac{X - iY}{X^2 + Y^2} \qquad (4.10)$$

where $$X + iY = \sqrt{(a^2 + b^2 - c^2 + 2ibc)}. \qquad (4.11)$$

By equating real and imaginary parts in (4.10), we deduce that

$$\int_0^\infty J_0(ax) \,.\, e^{-bx} \cos cx\,dx = \frac{X}{X^2 + Y^2}, \qquad (4.12)$$

$$\int_0^\infty J_0(ax) \,.\, e^{-bx} \sin cx\,dx = \frac{Y}{X^2 + Y^2}. \qquad (4.13)$$

From (4.11) we have

$$X^2 - Y^2 = a^2 + b^2 - c^2, \qquad (4.14)$$

$$XY = bc, \qquad (4.15)$$

and hence, by eliminating Y and X in turn, we find that X^2 and $-Y^2$ are the two roots of the equation in θ

$$\frac{a^2}{c^2 + \theta} + \frac{b^2}{\theta} = 1, \qquad (4.16)$$

and that

$$2X^2 = a^2 + b^2 - c^2 + \sqrt{\{(a^2 + b^2 - c^2)^2 + 4b^2c^2\}}, \qquad (4.17)$$

$$2Y^2 = -(a^2 + b^2 - c^2) + \sqrt{\{(a^2 + b^2 - c^2)^2 + 4b^2c^2\}}. \qquad (4.18)$$

Suppose $a > 0$, $c > 0$. Then, by following the continuous change in $X + iY$ when c increases from 0, we see from (4.11) that X and Y are both positive.

Now let $b \to 0$; then, if $a > c$, $X \to \sqrt{(a^2 - c^2)}$, $Y \to 0$; but, if $a < c$, $X \to 0$, $Y \to \sqrt{(c^2 - a^2)}$. Hence we find from (4.12), (4.13) respectively

$$\int_0^\infty J_0(ax) \,.\, \cos cx\, dx = 0, \qquad (a < c), \quad (4.19)$$

$$= \frac{1}{\sqrt{(a^2 - c^2)}}, \qquad (a > c)\,; \quad (4.20)$$

$$\int_0^\infty J_0(ax) \,.\, \sin cx\, dx = \frac{1}{\sqrt{(c^2 - a^2)}}, \qquad (a < c), \quad (4.21)$$

$$= 0, \qquad (a > c), \quad (4.22)$$

We have supposed $a > 0$, $c > 0$. If $a < 0$ or $c < 0$, we need only note that both integrals are even functions of a, and that the first is an even function of c and the second an odd function of c.

§ 57. We shall next prove that

$$\int_0^\infty J_0(ax) \,.\, e^{-bx} \,.\, \frac{\sin cx}{x} dx = \tan^{-1}\frac{c}{X} \quad . \quad (4.23)$$

where X is given by (4.17).

Proof. Since the infinite integral on the left of (4.12) is uniformly convergent with respect to c, we may integrate with respect to c under the integral sign from 0 to c. We thus find

$$\int_0^\infty J_0(ax) \,.\, e^{-bx} \,.\, \frac{\sin cx}{x} dx = \int_0^c \frac{Xdc}{X^2 + Y^2}. \quad (4.24)$$

Now from (4.14), (4.15),

$$XdX - YdY + cdc = 0,$$
$$YdX + XdY - bdc = 0,$$

and hence

$$\frac{dc}{X^2 + Y^2} = \frac{dX}{bY - cX} = \frac{Xdc - cdX}{X(X^2 + c^2)}, \quad . \quad (4.25)$$

from which follows

$$\int \frac{Xdc}{X^2 + Y^2} = \int \frac{Xdc - cdX}{X^2 + c^2} = \tan^{-1}\frac{c}{X}\,; \quad . \quad (4.26)$$

hence and from (4.24) follows (4.23), which was to be proved.

§ 58. Now suppose $a > 0$, $c > 0$, in (4.23), and let $b \to 0$. Then, as above, if $a > c$, $X \to \sqrt{(a^2 - c^2)}$, and

$$\tan^{-1}(c/X) \to \tan^{-1}\{c/\sqrt{(a^2 - c^2)}\} = \sin^{-1}(c/a);$$

but, if $a < c$, $X \to 0$, and $\tan^{-1}(c/X) \to \pi/2$.

Accordingly, we have

$$\int_0^\infty J_0(ax)\frac{\sin cx}{x}dx = \frac{\pi}{2}, \qquad (a < c) \quad . \quad (4.27)$$

$$= \sin^{-1}\frac{c}{a}, \quad (a > c). \quad . \quad (4.28)$$

If $c = 0$ the integral vanishes. If $a < 0$, or $c < 0$, we need only note that the integral is an even function of a, but an odd function of c.

Ex. Deduce the following well-known integrals as particular cases of the integrals in §§ 56-58 :—

$$\int_0^\infty e^{-bx}\cos cx\,dx = \frac{b}{b^2 + c^2}, \quad \int_0^\infty e^{-bx}\sin cx\,dx = \frac{c}{b^2 + c^2},$$

$$\int_0^\infty e^{-bx}\frac{\sin cx}{x}dx = \tan^{-1}\frac{c}{b},$$

$$\int_0^\infty \frac{\sin cx}{x}dx = \frac{\pi}{2}, \quad 0, \quad -\frac{\pi}{2}, \quad \text{according as } c >, \ =, \ < 0.$$

§ 59. *Electrostatic potential of an electrified disc.*

It was shown by Weber * that the potential of the electrostatic field caused by an electrified circular disc could be expressed as an integral of the same form as (4.23).

Let the disc be situated in the xy plane, with its centre at the origin and its axis along the z-axis ; let c be the radius of the disc, and Q the total charge of electricity upon it. Let V be the potential at any point due to the charge on the disc.

The differential equation satisfied by V in free space is Laplace's equation

$$\frac{\partial^2 V}{\partial x^2} + \frac{\partial^2 V}{\partial y^2} + \frac{\partial^2 V}{\partial z^2} = 0, \quad . \quad . \quad (4.29)$$

* Cf. Riemann-Weber : "Die Partiellen Differential-Gleichungen d. Math. Physik," I, 6th edn., 1919, p. 342.

which, in cylindrical co-ordinates r, θ, z, when V is independent of θ, as in the present problem, becomes

$$\frac{\partial^2 V}{\partial r^2} + \frac{1}{r}\frac{\partial V}{\partial r} + \frac{\partial^2 V}{\partial z^2} = 0. \quad . \quad . \quad (4.30)$$

This equation, by (3.17), has a solution of the form $V = e^{-\alpha z} J_0(\alpha r)$, where α is any constant, and it follows, by differentiation under the integral sign, that if $z > 0$

$$V = A\int_0^\infty e^{-\alpha z} J_0(\alpha r) f(\alpha) d\alpha \quad . \quad . \quad (4.31)$$

is also a solution, where A denotes any constant, and $f(\alpha)$ any function of α.

Now the charge distributes itself so as to make V = const. over the disc but not beyond it, and this condition is satisfied if we put $f(\alpha) = \dfrac{\sin c\alpha}{\alpha}$, for we then have

$$V = A\int_0^\infty e^{-\alpha z} J_0(\alpha r) \frac{\sin c\alpha}{\alpha} d\alpha, \quad . \quad . \quad (4.32)$$

which, by (4.27), (4.28), reduces when $z \to 0$ to $V = \frac{1}{2}\pi A$ if $r < c$, $V = A \sin^{-1}(c/r)$ if $r > c$.

It remains to determine the constant A in terms of the charge Q. Now, if σ is the surface density on the upper face of the disc, we have

$$4\pi\sigma = -\left(\frac{\partial V}{\partial z}\right)_{z=0} = A\int_0^\infty J_0(\alpha r) \sin c\alpha \, d\alpha = \frac{A}{\sqrt{(c^2 - r^2)}}$$

if $r < c$, by (4.21); and hence

$$\sigma = \frac{A}{4\pi\sqrt{(c^2 - r^2)}}. \quad . \quad . \quad (4.33)$$

Taking into account both faces of the disc, we have therefore

$$Q = 2\int_0^c \sigma \, . \, 2\pi r \, dr = A\int_0^c \frac{r \, dr}{\sqrt{(c^2 - r^2)}} = Ac. \quad (4.34)$$

Hence $A = Q/c$, and by (4.33), (4.32),

$$\sigma = \frac{Q}{4\pi c\sqrt{(c^2 - r^2)}}, \quad . \quad . \quad (4.35)$$

$$V = \frac{Q}{c}\int_0^\infty e^{-\alpha z} J_0(\alpha r) \frac{\sin c\alpha}{\alpha} d\alpha, \quad . \quad (4.36)$$

the constant potential over the disc being $\pi Q/2c$.

This gives V for positive values of z. For negative values of z we need only note that V is an even function of z.

§ 60. Further, by (4.23) it follows that

$$V = \frac{Q}{c}\tan^{-1}\frac{c}{X}, \quad . \quad . \quad . \quad (4.37)$$

where

$$2X^2 = r^2 + z^2 - c^2 + \sqrt{\{(r^2 + z^2 - c^2)^2 + 4c^2z^2\}} \quad (4.38)$$

Formulæ (4.33), (4.37) can be found by other methods (cf. Jeans : "Electricity and Magnetism," § 288).

Ex. Verify from (4.37) that $V \doteqdot \frac{Q}{\sqrt{(r^2 + z^2)}}$ when $\sqrt{(r^2 + z^2)}$ is large compared with c.

Examples VII

1. Show that

$$\int_0^\infty J_1(x)dx = 1.$$

2. Show that $J_0'(x) \to 0$ when $x \to \infty$, and by integrating the equation

$$J_0''(x) + \frac{1}{x}J_0'(x) + J_0(x) = 0$$

and using (4.7), show that

$$\int_0^\infty \frac{J_1(x)}{x}dx = 1.$$

3. If u, v, w denote the integrals

$$\int_0^\infty \frac{xJ_0(x)dx}{\sqrt{(a^2 + x^2)}}, \quad \int_0^\infty \frac{J_0'(x)dx}{\sqrt{(a^2 + x^2)}}, \quad \int_0^\infty \frac{xJ_0''(x)dx}{\sqrt{(a^2 + x^2)}},$$

respectively, show that

$$u + v + w = 0,$$

$$\frac{du}{da} + av + 1 = 0,$$

$$a\frac{dv}{da} - w = 0.$$

Deduce that, if $a > 0$,

$$\int_0^\infty \frac{xJ_0(x)dx}{\sqrt{(a^2 + x^2)}} = e^{-a}, \quad \int_0^\infty \frac{J_1(x)dx}{\sqrt{(a^2 + x^2)}} = \frac{1 - e^{-a}}{a}.$$

4. Show that

(i) $\int_0^a \int_0^{2\pi} \cos\left(\frac{r}{c} \cos \theta\right) . r\, dr\, d\theta = 2\pi ac J_1\left(\frac{a}{c}\right).$

(ii) $\int_0^a \int_0^{2\pi} \log \frac{a}{r} . \cos\left(\frac{r}{c} \cos \theta\right) . r\, dr\, d\theta = 2\pi c^2\left\{1 - J_0\left(\frac{a}{c}\right)\right\}.$

(iii) $\int_0^a \int_0^{2\pi} (a^2 - r^2) \cos\left(\frac{r}{c} \cos \theta\right) . r\, dr\, d\theta = 4\pi a^2 c^2 J_2\left(\frac{a}{c}\right).$

(iv) $\int_0^a \int_0^{2\pi} I_0\left(\frac{r}{b}\right) \cos\left(\frac{r}{c} \cos \theta\right) . r\, dr\, d\theta =$

$$\frac{2\pi abc}{b^2 + c^2}\left\{cJ_0\left(\frac{a}{c}\right) I_0'\left(\frac{a}{b}\right) - bI_0\left(\frac{a}{b}\right) J_0'\left(\frac{a}{c}\right)\right\}.$$

[As to the form of the result in (iii), see Exs. XI, 2, (i).]

5. If r, ρ, z denote the distances of a point from the origin, the z-axis, and the xy-plane respectively, show that if $z > 0$

$$\int_0^\infty e^{-zt} J_0(\rho t) dt = \frac{1}{r}.$$

Also, if R, R′ denote the distances of the point from $(0, 0, c)$, $(0, 0, -c)$ respectively, show that if $z > c$

$$\frac{R' - R}{2} = \int_0^\infty \frac{(zt + 1) \sinh ct - ct \cosh ct}{t^2} . e^{-zt} J_0(\rho t) dt.$$

6. Show that

$$I_0(x) = \frac{1}{2\pi}\int_0^{2\pi} e^{-x \sin \theta}\, d\theta = \frac{1}{2\pi}\int_0^{2\pi} e^{x \sin \theta}\, d\theta.$$

§ 61. *The Gamma-Function.*

In the next two chapters we shall need the elementary properties of the Gamma-function $\Gamma(n)$, which may be defined in the first place when n is real and positive by the integral

$$\Gamma(n) = \int_0^\infty e^{-x} x^{n-1} dx, \quad (n > 0); \qquad (4.39)$$

the condition $n > 0$ being necessary for the convergence of the integral at the lower limit.

In particular, when $n = 1$ we have

$$\Gamma(1) = \int_0^\infty e^{-x}dx = 1. \qquad (4.40)$$

Again, integrating by parts, we have

$$\Gamma(n) = \Big[-e^{-x}x^{n-1}\Big]_0^\infty + (n-1)\int_0^\infty e^{-x}x^{n-2}dx,$$

and therefore, if $n > 1$,

$$\Gamma(n) = (n-1)\Gamma(n-1), \qquad (4{\cdot}41)$$

and hence, when n is replaced by $n + 1$,

$$\Gamma(n+1) = n\Gamma(n). \qquad (4.42)$$

It follows by repeated application of this formula that, if n is a positive integer,

$$\Gamma(n+1) = n(n-1)(n-2)\ \ldots\ 3\,.\,2\,.\,1\ \Gamma(1),$$

that is, by (4.40),

$$\Gamma(n+1) = n\,! \qquad (4.43)$$

If we substitute x^2 for x in (4.39) we have also

$$\Gamma(n) = 2\int_0^\infty e^{-x^2}x^{2n-1}dx. \qquad (4.44)$$

§ 62. The integral

$$\int_0^{\frac{\pi}{2}} \cos^m\theta \sin^n\theta\, d\theta$$

can be expressed in terms of Gamma-functions.

We consider the double integral

$$u = \int_0^\infty\int_0^\infty e^{-x^2-y^2}x^{2m-1}y^{2n-1}dx\,dy$$

in two ways. Firstly we have, by (4.44),

$$u = \int_0^\infty e^{-x^2}x^{2m-1}dx\int_0^\infty e^{-y^2}y^{2n-1}dy = \tfrac{1}{4}\Gamma(m)\Gamma(n).$$

Secondly, by transforming to polar co-ordinates,

$$u = \int_0^{\frac{\pi}{2}}\int_0^\infty e^{-r^2}(r\cos\theta)^{2m-1}(r\sin\theta)^{2n-1}r\,dr\,d\theta$$

$$= \int_0^\infty e^{-r^2}r^{2m+2n-1}dr\int_0^{\frac{\pi}{2}}\cos^{2m-1}\theta\sin^{2n-1}\theta\,d\theta$$

$$= \tfrac{1}{2}\Gamma(m+n)\int_0^{\frac{\pi}{2}}\cos^{2m-1}\theta\sin^{2n-1}\theta\,d\theta,$$

by (4.44). Equating the two values of u thus found, we have, if $m > 0$, $n > 0$,

$$\int_0^{\frac{\pi}{2}} \cos^{2m-1}\theta \sin^{2n-1}\theta \, d\theta = \frac{\Gamma(m)\Gamma(n)}{2\Gamma(m+n)}. \qquad (4.45)$$

It follows that, if $m > -1$, $n > -1$,

$$\int_0^{\frac{\pi}{2}} \cos^m \theta \sin^n \theta \, d\theta = \frac{\Gamma\left(\frac{m+1}{2}\right)\Gamma\left(\frac{n+1}{2}\right)}{2\Gamma\left(\frac{m+n+2}{2}\right)}. \qquad (4.46)$$

In particular, putting $m = 0$, $n = 0$, we have

$$\frac{\pi}{2} = \int_0^{\frac{\pi}{2}} d\theta = \frac{\{\Gamma(\frac{1}{2})\}^2}{2\Gamma(1)} = \frac{\{\Gamma(\frac{1}{2})\}^2}{2},$$

and hence

$$\Gamma(\tfrac{1}{2}) = \sqrt{\pi}. \qquad (4.47)$$

Further, from (4.41), we have

$$\Gamma(\tfrac{3}{2}) = \tfrac{1}{2}\Gamma(\tfrac{1}{2}) = \tfrac{1}{2}\sqrt{\pi},$$

$$\Gamma(\tfrac{5}{2}) = \tfrac{3}{2}\Gamma(\tfrac{3}{2}) = \frac{1 \cdot 3}{2^2}\sqrt{\pi},$$

$$\Gamma(\tfrac{7}{2}) = \tfrac{5}{2}\Gamma(\tfrac{5}{2}) = \frac{1 \cdot 3 \cdot 5}{2^3}\sqrt{\pi},$$

and so on.

§ 63. Again, from (4.42), we have

$$\Gamma(n) = \frac{\Gamma(n+1)}{n}, \qquad (4.48)$$

from which it follows that $\Gamma(n) \to +\infty$ when $n \to +0$.

For our present purpose we may now suppose that (4.48) *defines* $\Gamma(n)$ firstly for values of n between -1 and 0, then for values between -2 and -1, then for values between -3 and -2, and so on; the Gamma-function will then have been defined for all real values of n; for example, by (4.48) and (4.47), we shall have

$$\Gamma(-\tfrac{1}{2}) = \frac{\Gamma(\frac{1}{2})}{-\frac{1}{2}} = -2\sqrt{\pi},$$

$$\Gamma(-\tfrac{3}{2}) = \frac{\Gamma(-\frac{1}{2})}{-\frac{3}{2}} = \frac{2^2}{1 \cdot 3}\sqrt{\pi},$$

and so on.

The graph of $\Gamma(n)$ for real values of n is indicated in Fig. 15. Note that $\Gamma(n) \to \pm \infty$ as n approaches a negative integer or zero.

§ 64. *The Beta-function.*

The Beta-function, $B(m, n)$, may be defined, in the first place for positive values of m and n, by the integral

$$B(m, n) = \int_0^1 x^{m-1}(1-x)^{n-1}\,dx. \quad . \quad . \quad (4.49)$$

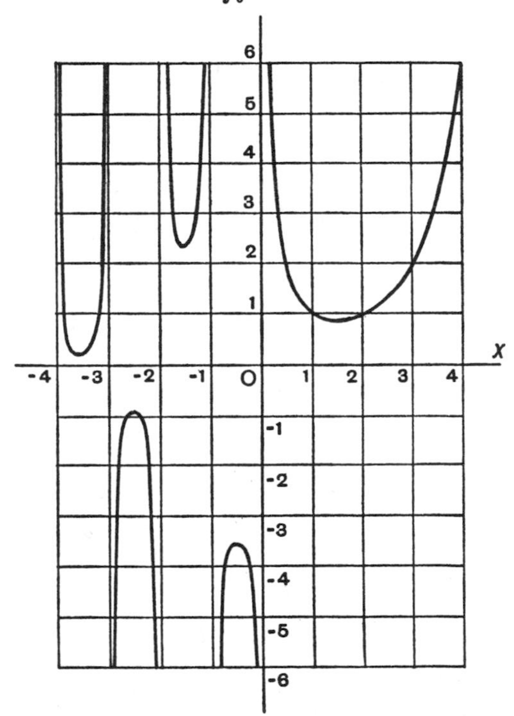

Fig. 15.

The conditions $m > 0$, $n > 0$ are necessary to ensure convergence of the integral at the lower and upper limits respectively.

The Beta-function can be expressed in terms of the Gamma-function. For, if we put $x = \cos^2 \theta$, we find

$$B(m, n) = 2\int_0^{\frac{\pi}{2}} \cos^{2m-1}\theta \sin^{2n-1}\theta\,d\theta, \quad . \quad . \quad (4.50)$$

and hence, by (4.45),

$$B(m, n) = \frac{\Gamma(m)\Gamma(n)}{\Gamma(m+n)}. \quad . \quad . \quad . \quad (4.51)$$

EXAMPLES VIII

1. Show that

(i) $\int_0^\infty e^{-ax}x^{n-1}dx = \frac{\Gamma(n)}{a^n}, \quad (n>0, \qquad a>0).$

(ii) $\int_0^\infty e^{-x^2}dx = \frac{\sqrt{\pi}}{2}.$

(iii) $\int_a^\infty e^{2ax-x^2}dx = \frac{\sqrt{\pi}}{2}e^{a^2}.$

(iv) $\int_0^\infty x^m e^{-x^n}dx = \frac{1}{n}\,\Gamma\left(\frac{m+1}{n}\right), \quad (m > -1, n > 0).$

2. Show that, if $n > -1$,

$$\int_0^{\frac{\pi}{2}} \sin^n\theta\,d\theta = \int_0^{\frac{\pi}{2}} \cos^n\theta\,d\theta = \frac{\sqrt{\pi}\,\Gamma\left(\frac{n+1}{2}\right)}{2\Gamma\left(\frac{n+2}{2}\right)}.$$

3. Show that, if $-1 < n < 1$,

$$\int_0^{\frac{\pi}{2}} \tan^n\theta\,d\theta = \frac{1}{2}\Gamma\left(\frac{1+n}{2}\right)\Gamma\left(\frac{1-n}{2}\right).$$

4. By evaluating the integral

$$\int_0^{\frac{\pi}{2}} \sin^{2n-1}\theta\cos^{2n-1}\theta\,d\theta,$$

in two ways, show that

$$\Gamma(n)\Gamma(n+\tfrac{1}{2}) = 2^{1-2n}\sqrt{\pi}\,\Gamma(2n).$$

5. Evaluate Lipschitz's integral

$$\int_0^\infty e^{-ax}J_0(bx)dx,$$

when $0 < b < a$, by using the expansion of $J_0(bx)$, and Ex. 1, (i).

6. Prove that, if $a > 0$,

$$\int_0^\infty e^{-ax}J_0(b\sqrt{x})dx = \frac{1}{a}e^{-b^2/4a}.$$

7. Show that, if $m > 0$, $n > 0$,

$$B(m, n) = \int_1^\infty \frac{(u-1)^{n-1}}{u^{m+n}}du = \int_0^\infty \frac{v^{n-1}}{(1+v)^{m+n}}dv.$$

§ 65. *Euler's constant in an integral form.*

As a preliminary to the next paragraph, we shall now obtain one form in which Euler's constant γ (§ 6) can be expressed by definite integrals. We have

$$\gamma = \lim_{n \to \infty}\left(1 + \frac{1}{2} + \frac{1}{3} + \dots + \frac{1}{n} - \log n\right)$$
$$= \lim_{n \to \infty}(s_n - \log n),$$

where
$$s_n = 1 + \frac{1}{2} + \frac{1}{3} + \dots + \frac{1}{n}.$$

Now we can write s_n in the form

$$s_n = \int_0^1 (1 + x + x^2 + \dots + x^{n-1})dx = \int_0^1 \frac{1 - x^n}{1 - x}dx.$$

Put
$$1 - x = \frac{t}{n}, \quad x = 1 - \frac{t}{n}, \quad dx = -\frac{dt}{n};$$

then

$$s_n = \int_0^n \left\{1 - \left(1 - \frac{t}{n}\right)^n\right\}\frac{dt}{t}$$
$$= \left(\int_0^1 + \int_1^n\right)\left\{1 - \left(1 - \frac{t}{n}\right)^n\right\}\frac{dt}{t}$$
$$= \int_0^1 \left\{1 - \left(1 - \frac{t}{n}\right)^n\right\}\frac{dt}{t} + \log n - \int_1^n \left(1 - \frac{t}{n}\right)^n \frac{dt}{t},$$

and hence

$$s_n - \log n = \int_0^1 \left\{1 - \left(1 - \frac{t}{n}\right)^n\right\}\frac{dt}{t} - \int_1^n \left(1 - \frac{t}{n}\right)^n \frac{dt}{t}.$$

Proceeding to the limit * when $n \to \infty$, we obtain

$$\gamma = \int_0^1 \frac{1 - e^{-t}}{t}dt - \int_1^\infty \frac{e^{-t}}{t}dt. \qquad (4.52)$$

§ 66. *The Integral* $\int_x^\infty \frac{e^{-t}}{t}dt$.

This integral will be required in § 75. We have, if $x > 0$,

$$\int_x^\infty \frac{e^{-t}}{t}dt = \int_x^1 \frac{e^{-t}}{t}dt + \int_1^\infty \frac{e^{-t}}{t}dt$$
$$= \int_x^1 \frac{1 - (1 - e^{-t})}{t}dt + \int_1^\infty \frac{e^{-t}}{t}dt$$
$$= -\log x - \int_x^1 \frac{1 - e^{-t}}{t}dt + \int_1^\infty \frac{e^{-t}}{t}dt$$
$$= -\log x - \left(\int_0^1 - \int_0^x\right)\frac{1 - e^{-t}}{t}dt + \int_1^\infty \frac{e^{-t}}{t}dt,$$

* See Bromwich: "Infinite Series," p. 459; Whittaker and Watson: "Modern Analysis," § 12.2.

and hence, by (4.52),

$$\int_x^\infty \frac{e^{-t}}{t}dt = -\log x - \gamma + \int_0^x \frac{1-e^{-t}}{t}dt$$

$$= -\log x - \gamma + \int_0^x \left(1 - \frac{t}{2\,!} + \frac{t^2}{3\,!} - \dots\right)dt$$

$$= -\log x - \gamma + x - \frac{x^2}{2\,.\,2\,!} + \frac{x^3}{3\,.\,3\,!} - \dots \quad (4.53)$$

In particular, when x is small,

$$\int_x^\infty \frac{e^{-t}}{t}dt \doteqdot -\log x - \gamma. \quad (4.54)$$

EXAMPLES IX

1. Show that

$$\gamma = 1 + (\tfrac{1}{2} + \log \tfrac{1}{2}) + (\tfrac{1}{3} + \log \tfrac{2}{3}) + (\tfrac{1}{4} + \log \tfrac{3}{4}) + \dots$$

2. By integrating both terms in (4.52) by parts, show that

$$\gamma = -\int_0^\infty e^{-t}\log t\,dt.$$

3. Show that

$$\Gamma'(1) = -\gamma.$$

4. Show that

$$\int_1^\infty \frac{e^{-xv}}{v}dv = -\log x - \gamma + x - \frac{x^2}{2\,.\,2\,!} + \frac{x^3}{3\,.\,3\,!} - \dots$$

CHAPTER V

ASYMPTOTIC EXPANSIONS

§ 67. *Hankel's definite integral for* $\mathrm{J}_0(x)$.

Returning to the integral (4.3), viz.

$$\mathrm{J}_0(x) = \frac{1}{\pi}\int_{-\frac{\pi}{2}}^{\frac{\pi}{2}} e^{ix \sin \theta} d\theta,$$

and making the substitution

$$t = \sin\theta, \quad d\theta = \frac{dt}{\sqrt{(1-t^2)}},$$

we obtain

$$\mathrm{J}_0(x) = \frac{1}{\pi}\int_{-1}^{1} \frac{e^{ixt}}{\sqrt{(1-t^2)}} dt. \qquad (5.1)$$

The real part of the integrand is an even function of t, the imaginary part is an odd function ; hence we have also

$$\mathrm{J}_0(x) = \frac{2}{\pi}\int_{0}^{1} \frac{\cos xt}{\sqrt{(1-t^2)}} dt. \qquad (5.2)$$

§ 68. We can at once verify that the integral on the right of (5.1) is a solution of Bessel's equation, by differentiating under the sign of integration. For, put

$$y = \int_{-1}^{1} \frac{e^{ixt}}{\sqrt{(1-t^2)}} dt,$$

then

$$x\frac{d^2y}{dx^2} + \frac{dy}{dx} + xy = x\left(\frac{d^2y}{dx^2} + y\right) + \frac{dy}{dx}$$

$$= \int_{-1}^{1}\left\{x\sqrt{(1-t^2)}e^{ixt} + \frac{ite^{ixt}}{\sqrt{(1-t^2)}}\right\}dt$$

$$= \int_{-1}^{1} \frac{d}{dt}\{-i\sqrt{(1-t^2)}e^{ixt}\}dt$$

$$= \Big[-i\sqrt{(1-t^2)}e^{ixt}\Big]_{-1}^{1} = 0,$$

as was to be verified.

§ 69. *Hankel's contour integral.*

Now consider the integral

$$y = \int_a^b \frac{e^{ixt}}{\sqrt{(1-t^2)}}dt \quad . \qquad . \qquad . \quad (5.3)$$

as a contour integral in the plane of the complex variable t, assuming for simplicity that x is positive. Then we find, as above,

$$x\frac{d^2y}{dx^2} + \frac{dy}{dx} + xy = \Big[-i\sqrt{(1-t^2)}e^{ixt}\Big]_a^b$$

and hence, if we put $a = \pm 1$, $b = i\eta$, and let $\eta \to +\infty$,

$$x\frac{d^2y}{dx^2} + \frac{dy}{dx} + xy = 0.$$

More generally, it is easy to see that we should still get this result if we put $a = \pm 1$, $b = \mathrm{R}(\cos\beta + i\sin\beta)$, and let $\mathrm{R} \to \infty$, provided $\sin\beta$ is positive, that is, provided $0 < \beta < \pi$.

It follows that y is a solution of Bessel's equation if the integration is carried out along any path joining either of the points $t = 1$, $t = -1$, to an infinitely distant point in the upper half of the t-plane. Thus, the path might be any straight line drawn from either of the points $t = 1$, $t = -1$, to infinity in the upper half of the t-plane. (Actually, in the limit, this line may coincide with the real axis from 1 to $+\infty$, or from -1 to $-\infty$, but this cannot be inferred by differentiation under the sign of integration, because the integrals obtained in this way are not then convergent.)

§ 70. Again, the function

$$\frac{e^{ixt}}{\sqrt{(1-t^2)}}$$

is a regular function of t at all points in the t-plane, except the branch points $t = 1$, $t = -1$, of the denominator. Consequently, by Cauchy's theory of contour integration, the value of the integral

$$\int \frac{e^{ixt}}{\sqrt{(1-t^2)}}dt, \quad . \quad . \quad . \quad (5.4)$$

taken round any simple closed contour which does not enclose either of those points, is zero.

§ 71. It is sometimes convenient to choose a contour part of which consists of an arc of a circle of infinite radius with its centre at the origin. The value of the integral (5.4), taken along any such arc in the upper half of the t-plane, vanishes when x is positive, by a well-known theorem in contour integration.*

§ 72. When the integral (5.4) is to be evaluated along any contour, it is necessary to pay attention to the continuity of $\sqrt{(1-t^2)}$, the denominator of the integrand. We shall assume that x is positive, and confine ourselves to contours in the upper half of the t-plane; accordingly, for the sake of following the way in which $\sqrt{(1-t^2)}$ varies, we indicate in Figs. 16·1, 16·2 the correspondence between the upper half of the t-plane and a plane on which $\sqrt{(1-t^2)}$ is represented, the branch points $t = 1$, $t = -1$, in the first of these planes being marked by indentations. The value of $\sqrt{(1-t^2)}$ at $t = 0$ is taken to be $+1$.

§ 73. *Use of the contour* HBDFH. *Hankel functions.*

First take the integral (5.4) round the closed contour HBDFH. The part of the integral corresponding to each

* Whittaker and Watson, § 6.222.

of the infinitesimal circular quadrants at H and B vanishes, and so does that corresponding to the infinitely distant arc DF of an infinite circle (§ 71). Consequently, since the value of the integral taken round this closed contour is zero (§ 70),

$$\left(\int_{-1}^{1} + \int_{1}^{1+i\infty} + \int_{-1+i\infty}^{-1}\right)\frac{e^{ixt}}{\sqrt{(1-t^2)}}dt = 0. \qquad . \quad (5.5)$$

Now, by (5.1), we have

$$\pi J_0(x) = \int_{-1}^{1}\frac{e^{ixt}}{\sqrt{(1-t^2)}}dt; \qquad . \quad . \quad (5.6)$$

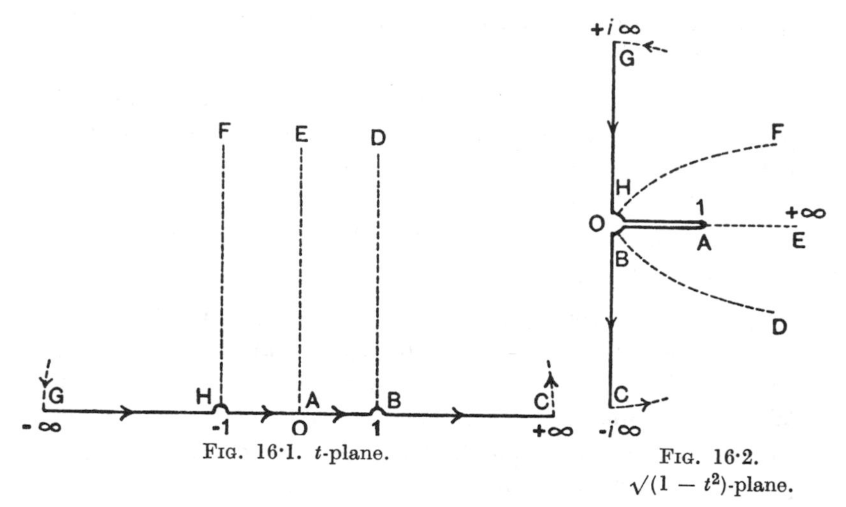

FIG. 16·1. t-plane.

FIG. 16·2. $\sqrt{(1-t^2)}$-plane.

and, in accordance with an accepted notation, we put

$$-\frac{\pi}{2}H_0^{(1)}(x) = \int_{1}^{1+i\infty}\frac{e^{ixt}}{\sqrt{(1-t^2)}}dt, \qquad . \quad (5.7)$$

$$-\frac{\pi}{2}H_0^{(2)}(x) = \int_{-1+i\infty}^{-1}\frac{e^{ixt}}{\sqrt{(1-t^2)}}dt; \qquad . \quad (5.8)$$

then, from (5.5), we have

$$J_0(x) = \tfrac{1}{2}\{H_0^{(1)}(x) + H_0^{(2)}(x)\}. \quad . \quad . \quad (5.9)$$

§ 74. The functions $H_0^{(1)}(x)$, $H_0^{(2)}(x)$ which are also solutions of Bessel's equation (§ 69), are called Hankel functions after Hankel, who wrote an outstanding memoir on Bessel functions.*

We shall next express these functions as integrals in which the range of integration is real. Along BD put

$$t = 1 + i\eta, \quad dt = id\eta\,;$$

then, paying attention to the continuity of $\sqrt{(1 - t^2)}$, (§ 72), we must write

$$\sqrt{(1 - t^2)} = \sqrt{(-\,2i\eta + \eta^2)} = \sqrt{2}e^{-\frac{1}{4}\pi i}\sqrt{(\eta + \tfrac{1}{2}i\eta^2)},$$

and hence, by (5.7),

$$-\,\frac{\pi}{2}\mathrm{H}_0^{(1)}(x) = \int_0^\infty \frac{e^{ix}e^{-x\eta}\,id\eta}{\sqrt{2}e^{-\frac{1}{4}\pi i}\sqrt{(\eta + \frac{1}{2}i\eta^2)}},$$

from which follows

$$\mathrm{H}_0^{(1)}(x) = \frac{\sqrt{2}}{\pi}e^{i(x - \frac{1}{4}\pi)}\int_0^\infty \frac{e^{-x\eta}}{\sqrt{(\eta + \frac{1}{2}i\eta^2)}}d\eta, \quad . \quad (5.10)$$

or further, putting $u = x\eta$,

$$\mathrm{H}_0^{(1)}(x) = \left(\frac{2}{\pi x}\right)^{\frac{1}{2}}\frac{e^{i(x - \frac{1}{4}\pi)}}{\sqrt{\pi}}\int_0^\infty \frac{e^{-u}}{\sqrt{u}}\left(1 + \frac{iu}{2x}\right)^{-\frac{1}{2}}du. \quad (5.11)$$

In the same kind of way we find from (5.8)

$$\mathrm{H}_0^{(2)}(x) = \left(\frac{2}{\pi x}\right)^{\frac{1}{2}}\frac{e^{-i(x - \frac{1}{4}\pi)}}{\sqrt{\pi}}\int_0^\infty \frac{e^{-u}}{\sqrt{u}}\left(1 - \frac{iu}{2x}\right)^{-\frac{1}{2}}du. \quad (5.12)$$

Thus, when x is positive, $\mathrm{H}_0^{(1)}(x)$, $\mathrm{H}_0^{(2)}(x)$ are a conjugate pair of complex numbers,† and $\mathrm{J}_0(x)$ is the real part of either, by (5.9). We shall prove that their imaginary parts are $i\mathrm{Y}_0(x)$, $-\,i\mathrm{Y}_0(x)$ respectively.

§ 75. *Use of the contour* ABDEA.

Take the integral (5.4) round the contour ABDEA. This gives

$$\left(\int_0^1 + \int_1^{1+i\infty} + \int_{i\infty}^0\right)\frac{e^{ixt}}{\sqrt{(1 - t^2)}}dt = 0. \quad . \quad (5.13)$$

Along AE we put

$$t = i\eta, \qquad dt = id\eta\,;$$
$$\sqrt{(1 - t^2)} = \sqrt{(1 + \eta^2)}\,;$$

* *Math. Annalen*, I, 1869.

† $\mathrm{H}_0^{(1)}(x)$, $\mathrm{H}_0^{(2)}(x)$ are not conjugate when x is not real.

then, having regard to § 67 and (5.7), we can write (5.13) in the form

$$\frac{\pi}{2}\mathrm{J}_0(x) + \int_0^1 \frac{i \sin xt}{\sqrt{(1-t^2)}}dt - \frac{\pi}{2}\mathrm{H}_0^{(1)}(x) + \int_\infty^0 \frac{e^{-x\eta}}{\sqrt{(1+\eta^2)}}id\eta = 0.$$

Putting

$$\mathrm{H}_0^{(1)}(x) = \mathrm{J}_0(x) + i\mathrm{Y}(x), \quad . \quad . \quad (5.14)$$

and separating the imaginary parts, we find

$$\frac{\pi}{2}\mathrm{Y}(x) = -\int_0^\infty \frac{e^{-x\eta}}{\sqrt{(1+\eta^2)}}d\eta + \int_0^1 \frac{\sin xt}{\sqrt{(1-t^2)}}dt,$$

which we may also write, with $\eta = \sinh u$, $t = \sin\theta$, $v = e^u$,

$$\frac{\pi}{2}\mathrm{Y}(x) = -\int_0^\infty e^{-x \sinh u}\, du + \int_0^{\frac{\pi}{2}} \sin(x \sin\theta)d\theta$$

$$= -\int_1^\infty e^{-\frac{x}{2}\left(v - \frac{1}{v}\right)}\frac{dv}{v} + \int_0^{\frac{\pi}{2}} \sin(x\sin\theta)d\theta.$$

The first integral on the right may be expanded in the form

$$-\int_1^\infty \frac{e^{-\frac{1}{2}xv}}{v}\left\{1 + \frac{x}{2v} + \frac{1}{2\,!}\left(\frac{x}{2v}\right)^2 + \ldots\right\}dv.$$

Now, suppose that x is small. Then, by Exs. IX, 4, the first term of this expansion is approximately equal to

$$\log(\tfrac{1}{2}x) + \gamma, \quad \text{or} \quad \log x - (\log 2 - \gamma).$$

The remaining terms are small in comparison; for example, since $e^{-\frac{1}{2}xv} < 1$, the numerical value of the second term is less than

$$\int_1^\infty \frac{x}{2v^2}dv, = \frac{x}{2}.$$

Again, since $\sin(x\sin\theta) < x\sin\theta$, the value of the second integral in the above expression for $\frac{1}{2}\pi\mathrm{Y}(x)$ is less than

$$\int_0^{\frac{\pi}{2}} x\sin\theta\, d\theta, = x.$$

It follows that, when x is small,

$$Y(x) = \frac{2}{\pi}\{\log x - (\log 2 - \gamma) \ldots\}. \quad . \quad (5.15)$$

Now, since $J_0(x)$ and $H_0^{(1)}(x)$ are solutions of Bessel's equation (§ 69), it follows from (5.14) that $Y(x)$ is also a solution, and hence that

$$Y(x) = AJ_0(x) + BY_0(x),$$

where A, B are certain constants. Then, by (5.15) and (1.16), when x is small,

$$\frac{2}{\pi}\{\log x - (\log 2 - \gamma) \ldots\}$$
$$= A(1 - \ldots) + B\frac{2}{\pi}\{\log x - (\log 2 - \gamma) \ldots\}.$$

Equating coefficients of $\log x$, and then the terms independent of x, we find $B = 1$, $A = 0$, and hence

$$Y(x) \equiv Y_0(x).$$

Consequently, by (5.9) and (5.14), we have now shown that

$$H_0^{(1)}(x) = J_0(x) + iY_0(x), \quad . \quad . \quad (5.16)$$
$$H_0^{(2)}(x) = J_0(x) - iY_0(x). \quad . \quad . \quad (5.17)$$

Also

$$J_0(x) = \tfrac{1}{2}\{H_0^{(1)}(x) + H_0^{(2)}(x)\}, \quad . \quad . \quad (5.18)$$
$$Y_0(x) = -\tfrac{1}{2}i\{H_0^{(1)}(x) - H_0^{(2)}(x)\}. \quad . \quad (5.19)$$

§ 76. *Use of the contour* DBCD.

If x is positive, the integral (5.4), taken along the part of an infinite circle that bounds the positive quadrant, vanishes (§ 71). Hence, if we evaluate the integral round the closed contour DBCD, part of which consists of an infinite circular quadrant, we get

$$\left(\int_{1+i\infty}^{1} + \int_{1}^{\infty}\right)\frac{e^{ixt}}{\sqrt{(1-t^2)}}dt = 0,$$

that is, by (5.7)

$$\frac{\pi}{2}H_0^{(1)}(x) + \int_1^{\infty}\frac{e^{ixt}}{\sqrt{(1-t^2)}}dt = 0.$$

Now, along BC (see Fig. 16·2),

$$\sqrt{(1-t^2)} = -i\sqrt{(t^2-1)},$$

and therefore

$$\frac{\pi}{2}\mathrm{H}_0^{(1)}(x) = -i\int_1^\infty \frac{e^{ixt}}{\sqrt{(t^2-1)}}dt, \qquad (5.20)$$

from which follows, on equating real and imaginary parts,

$$\mathrm{J}_0(x) = \frac{2}{\pi}\int_1^\infty \frac{\sin xt}{\sqrt{(t^2-1)}}dt, \qquad (5.21)$$

$$\mathrm{Y}_0(x) = -\frac{2}{\pi}\int_1^\infty \frac{\cos xt}{\sqrt{(t^2-1)}}dt. \qquad (5.22)$$

Ex. 1. Obtain *Mehler's* integral forms:

$$\mathrm{J}_0(x) = \frac{2}{\pi}\int_0^\infty \sin(x\cosh u)du,$$

$$\mathrm{Y}_0(x) = -\frac{2}{\pi}\int_0^\infty \cos(x\cosh u)du.$$

Ex. 2. *Struve's* function of zero order, $\mathbf{H}_0(x)$, may be defined by

$$\int_0^1 \frac{e^{ixt}}{\sqrt{(1-t^2)}}dt = \frac{\pi}{2}\{\mathrm{J}_0(x) + i\mathbf{H}_0(x)\}.$$

Show that

$$\frac{\pi}{2}\mathbf{H}_0(x) = \frac{x}{1^2} - \frac{x^3}{1^2 \cdot 3^2} + \frac{x^5}{1^2 \cdot 3^2 \cdot 5^2} - \dots \qquad (5.23)$$

Also, by taking the integral

$$\int \frac{e^{ixt}}{\sqrt{(1-t^2)}}dt$$

round the contour ABDEA show that

$$\frac{2}{\pi}\int_0^\infty \frac{e^{-x\eta}}{\sqrt{(1+\eta^2)}}d\eta = \mathbf{H}_0(x) - \mathrm{Y}_0(x).$$

§ 77. *Asymptotic power-series and asymptotic expansions.* We shall digress to explain the meaning of an asymptotic power-series and an asymptotic expansion.

Definition 1.—An expansion of a function $f(x)$ in the form

$$f(x) = a_0 + \frac{a_1}{x} + \frac{a_2}{x^2} + \frac{a_3}{x^3} + \dots \qquad (5.24)$$

will be called an *asymptotic power-series* for $f(x)$, if it has the property that, for every fixed value of n,

$$f(x) = a_0 + \frac{a_1}{x} + \frac{a_2}{x^2} + \ldots + \frac{a_{n-1}}{x^{n-1}} + \frac{a_n + \epsilon_n(x)}{x^n}, \quad (5.25)$$

where $\epsilon_n(x) \to 0$ when $x \to \infty$.

It follows from the definition that :—

I. A convergent power-series of the form (5.24) is a particular case of an asymptotic power-series.

II. The sum $S_n(x)$ of the first n terms of an asymptotic power-series for $f(x)$ is an approximation to $f(x)$, in the sense that the difference $f(x) - S_n(x)$ is as small as we please compared with $1/x^{n-1}$ when x is large enough.

Proof. Put

$$R_n(x) = f(x) - S_n(x) = \frac{a_n + \epsilon_n(x)}{x^n};$$

then

$$\frac{R_n(x)}{1/x^{n-1}} = \frac{a_n + \epsilon_n(x)}{x},$$

which is as small as we please when x is large enough, since $\epsilon_n(x) \to 0$ by the definition.

COR. The difference $f(x) - S_n(x)$ is as small as we please compared with the last non-zero term in $S_n(x)$, when x is large enough.

III. The sum of the asymptotic power-series of two functions $f(x)$, $g(x)$ is the asymptotic power-series of their sum $f(x) + g(x)$.

IV. A given function $f(x)$ cannot have two different asymptotic power-series for the same range of values of x.

To prove this, suppose two different series possible, and consider their difference.

V. On the other hand, two different functions can have the same asymptotic power-series.

For example, it follows from the definition, and by remembering that $\lim\limits_{x \to \infty} (x^n e^{-x}) = 0$, that the asymptotic power-series for e^{-x} is $0 + 0 + 0 + \ldots$, $(x > 0)$. Consequently, $f(x) + e^{-x}$ has the same asymptotic power-series as $f(x)$, for $x > 0$.

VI. A function may have no asymptotic power-series.

Such a function is e^x, $(x > 0)$, since $e^x \to +\infty$ when $x \to +\infty$.

VII. The product of the asymptotic power-series of two functions $f(x)$, $g(x)$ is the asymptotic power-series of their product $f(x)g(x)$.

In particular, since the asymptotic power-series of e^{-x} is $0 + 0 + 0 + \ldots$, it follows that, if $f(x)$ has an asymptotic power-series, then the asymptotic power-series of the product $e^{-x}.f(x)$ is $0 + 0 + 0 + \ldots$, $(x > 0)$.

VIII. Let $f(x)$ be a function whose derivatives of all orders exist at $x = 0$. Then Maclaurin's series

$$f(x) = f(0) + xf'(0) + \frac{x^2}{2!}f''(0) + \frac{x^3}{3!}f'''(0) + \ldots$$

whether it converges or not, has the property that, for every fixed value of n,

$$f(x) = f(0) + xf'(0) + \ldots + \frac{x^{n-1}}{(n-1)!}f^{(n-1)}(0) + \frac{x^n}{n!}\{f^{(n)}(0) + \eta_n(x)\}$$

where $\eta_n(x) \to 0$ when $x \to 0$.

By writing $1/x$ for x it follows that, if $f^{(n)}(0)$ exists for all values for n, the function $f(1/x)$ is represented asymptotically by the power-series

$$f\left(\frac{1}{x}\right) = f(0) + \frac{f'(0)}{x} + \frac{f''(0)}{2!\,x^2} + \frac{f'''(0)}{3!\,x^3} + \ldots \qquad (5.26)$$

for large values of x.

§ 78. *Definition* 2.—If the asymptotic power-series of the quotient $f(x)/g(x)$ is given by

$$\frac{f(x)}{g(x)} = a_0 + \frac{a_1}{x} + \frac{a_2}{x^2} + \frac{a_3}{x^3} + \ldots$$

we shall say that

$$f(x) = g(x)\left(a_0 + \frac{a_1}{x} + \frac{a_2}{x^2} + \frac{a_3}{x^3} + \ldots\right)$$

is an *asymptotic expansion* of $f(x)$.

We shall also say that the sum of the asymptotic expansions of two functions is an asymptotic expansion of their sum.

§ 79. *Asymptotic expansions of the Hankel functions.*

Now write (5.11) in the form

$$\mathrm{H}_0^{(1)}(x) = \left(\frac{2}{\pi x}\right)^{\frac{1}{2}} \frac{e^{i(x-\frac{1}{4}\pi)}}{\sqrt{\pi}} f\left(\frac{1}{x}\right) \quad . \quad . \quad (5.27)$$

where

$$f(x) = \int_0^\infty \frac{e^{-u}}{\sqrt{u}}\left(1 + \frac{iux}{2}\right)^{-\frac{1}{2}} du.$$

Differentiating n times under the integral sign and putting $x = 0$, we find

$$f^{(n)}(0) = (-)^n \frac{1}{2}\frac{3}{2}\frac{5}{2} \cdots \frac{2n-1}{2} \cdot \frac{i^n}{2^n}\int_0^\infty e^{-u}\, u^{n-\frac{1}{2}}\, du,$$

and hence, by §§ 61, 62,

$$f^{(n)}(0) = (-)^n \frac{1 \,.\, 3 \,.\, 5 \,.\,.\,.\, (2n-1)}{2^n} \cdot \frac{i^n}{2^n} \cdot \Gamma(n + \tfrac{1}{2})$$

$$= (-)^n\, 1^2 \,.\, 3^2 \,.\, 5^2 \,.\,.\,.\, (2n-1)^2 \,.\, \frac{i^n\sqrt{\pi}}{8^n}.$$

Thus, $f^{(n)}(0)$ exists for all values of n, and it follows, by (5.26), that $f(1/x)$ is represented for large values of x by the asymptotic power-series

$$\sqrt{\pi}\left\{1 - \frac{i}{8x} + \frac{1^2 \,.\, 3^2}{2\,!}\frac{i^2}{(8x)^2} - \frac{1^2 \,.\, 3^2 \,.\, 5^2}{3\,!}\frac{i^3}{(8x)^3} + \,.\,.\,.\right\}$$

Substituting this series for $f(1/x)$ in (5.27), we obtain

$$\mathrm{H}_0^{(1)}(x) = \left(\frac{2}{\pi x}\right)^{\frac{1}{2}} e^{i(x-\frac{1}{4}\pi)}\left\{1 - \frac{i}{8x} + \frac{1^2 \,.\, 3^2}{2\,!}\frac{i^2}{(8x)^2} - \frac{1^2 \,.\, 3^2 \,.\, 5^2}{3\,!}\frac{i^3}{(8x)^3} + \,.\,.\,.\right\};$$

and, replacing i by $-i$, we have also

$$H_0^{(2)}(x) = \left(\frac{2}{\pi x}\right)^{\frac{1}{2}} e^{-i(x-\frac{1}{4}\pi)}\left\{1 + \frac{i}{8x} + \frac{1^2 . 3^2}{2!}\frac{i^2}{(8x)^2} + \frac{1^2 . 3^2 . 5^2}{3!}\frac{i^3}{(8x)^3} + \ldots\right\}$$

These are asymptotic expansions of the Hankel functions in accordance with the definition of § 78. They have been derived on the assumption that x is positive (§ 69), but it can be proved * that they remain true when x is replaced by $z, = |z| e^{i\theta}$, provided that $-\pi < \theta < 2\pi$ in the case of $H_0^{(1)}(z)$, and that $-2\pi < \theta < \pi$ in the case of $H_0^{(2)}(z)$.

Note that the series in brackets do not converge for any value of x.

§ 80. *Asymptotic expansions of* $J_0(x)$ *and* $Y_0(x)$.

If we now put

$$P = 1 - \frac{1^2 . 3^2}{2!}\frac{1}{(8x)^2} + \frac{1^2 . 3^2 . 5^2 . 7^2}{4!}\frac{1}{(8x)^4} - \ldots \quad (5.28)$$

$$Q = \frac{1}{8x} - \frac{1^2 . 3^2 . 5^2}{3!}\frac{1}{(8x)^3} + \ldots, \quad (5.29)$$

we can write the asymptotic expansions of the Hankel functions in the form

$$H_0^{(1)}(x) = \left(\frac{2}{\pi x}\right)^{\frac{1}{2}} e^{i(x-\frac{1}{4}\pi)}(P - iQ), \quad (5.30)$$

$$H_0^{(2)}(x) = \left(\frac{2}{\pi x}\right)^{\frac{1}{2}} e^{-i(x-\frac{1}{4}\pi)}(P + iQ), \quad (5.31)$$

and hence, by (5.16),

$$J_0(x) = \left(\frac{2}{\pi x}\right)^{\frac{1}{2}}\left\{P \cos\left(x - \frac{\pi}{4}\right) + Q \sin\left(x - \frac{\pi}{4}\right)\right\}, \quad (5.32)$$

$$Y_0(x) = \left(\frac{2}{\pi x}\right)^{\frac{1}{2}}\left\{P \sin\left(x - \frac{\pi}{4}\right) - Q \cos\left(x - \frac{\pi}{4}\right)\right\} \quad (5.33)$$

which are the asymptotic expansions of the standard Bessel functions of zero order, in accordance with the second definition in § 78.

* Watson, p 198.

The following deductions may be made :—

I. $$\mathrm{J}_0(x) = \left(\frac{2}{\pi x}\right)^{\frac{1}{2}}\left\{\cos\left(x - \frac{\pi}{4}\right) + p(x)\right\}, \qquad (5.34)$$

$$\mathrm{Y}_0(x) = \left(\frac{2}{\pi x}\right)^{\frac{1}{2}}\left\{\sin\left(x - \frac{\pi}{4}\right) + q(x)\right\}, \qquad (5.35)$$

$$\mathrm{J}_0'(x) = -\left(\frac{2}{\pi x}\right)^{\frac{1}{2}}\left\{\sin\left(x - \frac{\pi}{4}\right) + r(x)\right\}, \qquad (5.36)$$

where $p(x)$, $q(x)$, $r(x)$ all $\to 0$, when $x \to +\infty$.*

II. The positive roots of the equation $\mathrm{J}_0(x) = 0$ are given approximately by

$$\alpha_s = (s - \tfrac{1}{4})\pi. \qquad (5.37)$$

This approximation is fairly good even for $s = 1, 2, 3, \ldots$

§ 81. *Asymptotic expansion of* $\mathrm{I}_0(x)$ *and* $\mathrm{K}_0(x)$.

The asymptotic expansions of the modified Bessel functions of zero order, $\mathrm{I}_0(x)$, $\mathrm{K}_0(x)$, may be deduced from the formulae

$$\mathrm{I}_0(x) = \mathrm{J}_0(ix) = \tfrac{1}{2}\{\mathrm{H}_0^{(1)}(ix) + \mathrm{H}_0^{(2)}(ix)\}, \qquad (5.38)$$

$$\mathrm{K}_0(x) = \frac{\pi i}{2}\mathrm{H}_0^{(1)}(ix), \qquad (5.39)$$

the second of which will be proved below (§ 82).

Assuming that the asymptotic expansions of $\mathrm{H}_0^{(1)}(x)$, $\mathrm{H}_0^{(2)}(x)$ hold good when x is replaced by ix (§ 79), we have

$$\mathrm{H}_0^{(1)}(ix) = -i\left(\frac{2}{\pi x}\right)^{\frac{1}{2}}e^{-x}\left\{1 - \frac{1}{8x} + \frac{1^2 . 3^2}{2\,!}\frac{1}{(8x)^2} - \ldots\right\} \quad (5.40)$$

$$\mathrm{H}_0^{(2)}(ix) = \left(\frac{2}{\pi x}\right)^{\frac{1}{2}}e^{x}\left\{1 + \frac{1}{8x} + \frac{1^2 . 3^2}{2\,!}\frac{1}{(8x)^2} + \ldots\right\} \quad (5.41)$$

From (5.40) and § 77, VII, it follows that, if $x > 0$, the asymptotic power-series of $\sqrt{x}\mathrm{H}_0^{(1)}(ix)$ is $0 + 0 + 0 + \ldots$ and hence further that

$$e^{-x}\sqrt{x}\mathrm{H}_0^{(1)}(ix) = 0 + 0 + 0 + \ldots; \qquad (5.42)$$

* As regards $\mathrm{J}_0'(x)$, an asymptotic power-series deduced from a Maclaurin series, as in § 78, VIII, can always be differentiated.

also, from (5.41), $e^{-x}\sqrt{x}H_0^{(2)}(ix)$ has the asymptotic power-series,

$$e^{-x}\sqrt{x}H_0^{(2)}(ix) = \left(\frac{2}{\pi}\right)^{\frac{1}{2}}\left\{1 + \frac{1}{8x} + \frac{1^2 \cdot 3^2}{2\,!}\frac{1}{(8x)^2} + \ldots\right\} \quad (5.43)$$

Remembering § 77, III, we have by addition and (5.38) the asymptotic power-series

$$2e^{-x}\sqrt{x}I_0(x) = \left(\frac{2}{\pi}\right)^{\frac{1}{2}}\left\{1 + \frac{1}{8x} + \frac{1^2 \cdot 3^2}{2\,!}\frac{1}{(8x)^2} + \ldots\right\},$$

and hence the asymptotic expansion

$$I_0(x) = \frac{e^x}{\sqrt{(2\pi x)}}\left\{1 + \frac{1}{8x} + \frac{1^2 \cdot 3^2}{2\,!}\frac{1}{(8x)^2} + \ldots\right\}. \quad (5.44)$$

Again, from (5.39) and (5.40), the asymptotic expansion of $K_0(x)$ is given by

$$K_0(x) = \left(\frac{\pi}{2x}\right)^{\frac{1}{2}}e^{-x}\left\{1 - \frac{1}{8x} + \frac{1^2 \cdot 3^2}{2\,!}\frac{1}{(8x)^2} - \ldots\right\}. \quad (5.45)$$

Note.—From the integral (5.20) we conclude that $H_0^{(1)}(ix)$ is a pure imaginary, and with this conclusion (5.40) is consistent. But (5.41) appears to imply that $H_0^{(2)}(ix)$ is real, and that this is not true follows from (5.38). Actually, from (5.41) we can only deduce that, in the asymptotic power-series (5.43) of $e^{-x}\sqrt{x}H_0^{(2)}(ix)$, when $x > 0$, the imaginary part is of the form $0 + 0 + 0 + \ldots$, and this is confirmed by (5.42), since the imaginary part of $H_0^{(2)}(ix)$ is $-H_0^{(1)}(ix)$, as we see from (5.38).

§ 82. It remains to prove (5.39).

Proof. Since $H_0^{(1)}(x)$ is a solution of Bessel's equation, it follows that $H_0^{(1)}(ix)$ is a solution of the modified equation (3.1), and hence, by (3.6), that

$$\frac{\pi i}{2}H_0^{(1)}(ix) = AI_0(x) + BK_0(x), \quad . \quad . \quad (5.46)$$

where A, B are certain constants.

Now, from (5.20) we have

$$\frac{\pi i}{2}H_0^{(1)}(ix) = \int_1^\infty \frac{e^{-xt}}{\sqrt{(t^2-1)}}dt \quad . \quad . \quad (5.47)$$

$$= \int_0^\infty e^{-x\cosh u}\,du$$

$$= \int_1^\infty e^{-\frac{x}{2}\left(v+\frac{1}{v}\right)}\frac{dv}{v},$$

and it follows, as in § 75, that when x is small

$$\frac{\pi i}{2}H_0^{(1)}(ix) = -\log x + (\log 2 - \gamma)\ .\ .\ .$$

Substituting in (5.46), and using (3.5), we have, when x is small,

$$-\log x + (\log 2 - \gamma)\ .\ .\ .$$
$$= A(1 + \ .\ .\ .) + B\{-\log x + (\log 2 - \gamma)\ .\ .\ .\}.$$

Equating coefficients of $\log x$, and then the terms independent of x, we find $B = 1$, $A = 0$. Consequently, putting these values in (5.46), we have

$$K_0(x) = \frac{\pi i}{2}H_0^{(1)}(ix), \quad . \quad . \quad (5.48)$$

as was to be proved.

Ex. 1. Obtain the asymptotic expansion of $K_0(x)$ directly from the integral (5.47), which, with (5.39), gives

$$K_0(x) = \int_1^\infty \frac{e^{-xt}}{\sqrt{(t^2-1)}}dt.$$

[Begin by putting $t = 1 + \frac{u}{x}$.]

Ex. 2. Calculate $J_0(x)$ from (5.32) when $x = 10$, and compare the result with that tabulated at the end of the book.

Ex. 3. From (5.44) obtain the asymptotic expansions of ber x and bei x (see § 51) :—

$$\text{ber}\,x = \frac{e^{x/\sqrt{2}}}{\sqrt{(2\pi x)}}\left\{P\cos\left(\frac{x}{\sqrt{2}} - \frac{\pi}{8}\right) + Q\sin\left(\frac{x}{\sqrt{2}} - \frac{\pi}{8}\right)\right\},$$

$$\text{bei}\,x = \frac{e^{x/\sqrt{2}}}{\sqrt{(2\pi x)}}\left\{P\sin\left(\frac{x}{\sqrt{2}} - \frac{\pi}{8}\right) - Q\cos\left(\frac{x}{\sqrt{2}} - \frac{\pi}{8}\right)\right\},$$

where

$$P = 1 + \frac{1}{8x}\cos\frac{\pi}{4} + \frac{1^2 \cdot 3^2}{2\,!(8x)^2}\cos\frac{2\pi}{4} + \ldots$$

$$Q = \frac{1}{8x}\sin\frac{\pi}{4} + \frac{1^2 \cdot 3^2}{2\,!(8x)^2}\sin\frac{2\pi}{4} + \ldots$$

Ex. 4. Show that the asymptotic expansion of Struve's function $\mathsf{H}_0(x)$ can be written

$$\mathsf{H}_0(x) = Y_0(x) + \frac{2}{\pi x}\left(1 - \frac{1}{x^2} + \frac{1^2 \cdot 3^2}{x^4} - \frac{1^2 \cdot 3^2 \cdot 5^2}{x^6} + \ldots\right).$$

[See § 76, Ex. 2.]

Ex. 5. Show that (see e.g. Hardy: "Pure Mathematics," 7th edn., § 167)

$$(1 + h)^{-\frac{1}{2}} = S_n + R_n,$$

where

$$S_n = 1 - \tfrac{1}{2}h + \frac{\frac{1}{2} \cdot \frac{3}{2}}{1 \cdot 2}h^2 - \frac{\frac{1}{2} \cdot \frac{3}{2} \cdot \frac{5}{2}}{1 \cdot 2 \cdot 3}h^3 + \ldots \text{ to } n \text{ terms},$$

$$R_n = (-)^n \frac{\frac{1}{2} \cdot \frac{3}{2} \cdot \frac{5}{2} \ldots (n - \frac{1}{2})}{(n - 1)\,!}h^n \int_0^1 (1 - t)^{n-1}(1 + ht)^{-\frac{1}{2}-n}dt.$$

If the real part of h is positive or zero, show further that

$$|\,R_n\,| < \frac{\frac{1}{2} \cdot \frac{3}{2} \cdot \frac{5}{2} \ldots (n - \frac{1}{2})}{n\,!}\,|\,h\,|^n.$$

Deduce that, in the asymptotic power-series for $f(1/x)$ in § 79, the modulus of the difference between $f(1/x)$ and the sum of the first n terms of the series is less than the modulus of the $(n + 1)$th term for every positive value of x.

CHAPTER VI

BESSEL FUNCTIONS OF ANY REAL ORDER

§ 83. *Bessel functions of any order.*

We can define $J_n(x)$, when n is any integer, to be the coefficient of t^n in the expansion of the function

$$e^{\frac{x}{2}\left(t-\frac{1}{t}\right)}, \quad . \quad . \quad . \quad . \qquad (6.1)$$

in ascending and descending powers of t. Now,

$$e^{\frac{xt}{2}} = 1 + \frac{xt}{2} + \frac{1}{2\,!}\left(\frac{xt}{2}\right)^2 + \frac{1}{3\,!}\left(\frac{xt}{2}\right)^3 + \ldots$$

$$e^{-\frac{x}{2t}} = 1 - \frac{x}{2t} + \frac{1}{2\,!}\left(\frac{x}{2t}\right)^2 - \frac{1}{3\,!}\left(\frac{x}{2t}\right)^3 + \ldots$$

Consequently, picking out the coefficient of t^n in the product of these two series, we find by the above definition, if n is a positive integer, that

$$J_n(x) = \frac{x^n}{2^n n\,!}\left(1 - \frac{x^2}{2\,.\,2n+2} + \frac{x^4}{2\,.\,4\,.\,2n+2\,.\,2n+4} - \ldots\right), \qquad (6.2)$$

and that

$$J_{-n}(x) = \frac{(-)^n x^n}{2^n n\,!}\left(1 - \frac{x^2}{2\,.\,2n+2} + \frac{x^4}{2\,.\,4\,.\,2n+2\,.\,2n+4} - \ldots\right), \qquad (6.3)$$

The first of these agrees with the original definition of $J_n(x)$ in § 2 ; the second can be written

$$J_{-n}(x) = (-)^n J_n(x). \quad . \quad . \qquad (6.4)$$

The graphs of $J_0(x)$, $J_1(x)$ have been given in Fig. 1. Those of $J_2(x)$, $J_4(x)$, $J_6(x)$ are shown in Fig. 17.

§ 84. If n is not an integer, $J_n(x)$ can be defined by the series

$$J_n(x) = \frac{x^n}{2^n\Gamma(n+1)}\left(1 - \frac{x^2}{2 \,.\, 2n+2} + \frac{x^4}{2 \,.\, 4 \,.\, 2n+2 \,.\, 2n+4} - \ldots\right), \qquad (6\cdot5)$$

where Γ denotes the Gamma-function (§ 61).

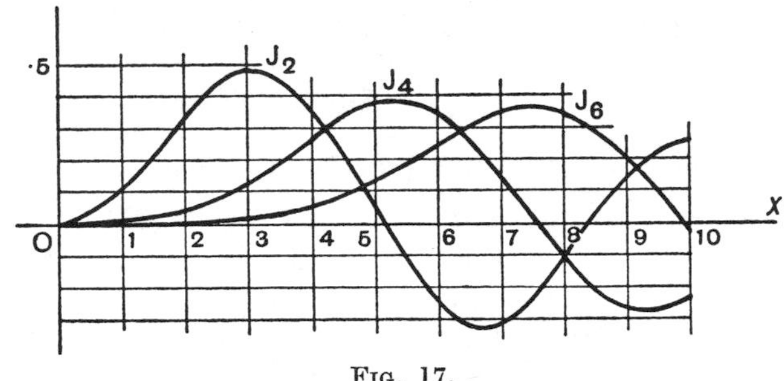

Fig. 17.

In particular, by putting $n = \frac{1}{2}$, $n = -\frac{1}{2}$ in turn, we find

$$J_{\frac{1}{2}}(x) = \left(\frac{2}{\pi x}\right)^{\frac{1}{2}} \sin x, \quad . \quad . \quad . \quad (6.6)$$

$$J_{-\frac{1}{2}}(x) = \left(\frac{2}{\pi x}\right)^{\frac{1}{2}} \cos x. \quad . \quad . \quad . \quad (6.7)$$

§ 85. The *modified Bessel function* $I_n(x)$ is defined by the series

$$I_n(x) = \frac{x^n}{2^n\Gamma(n+1)}\left(1 + \frac{x^2}{2 \,.\, 2n+2} + \frac{x^4}{2 \,.\, 4 \,.\, 2n+2 \,.\, 2n+4} + \ldots\right) \qquad (6.8)$$

[cf. (3.3)]. We note that

$$I_n(x) = i^{-n}J_n(ix). \quad . \quad . \quad . \quad (6.9)$$

In particular, we find

$$I_{\frac{1}{2}}(x) = \left(\frac{2}{\pi x}\right)^{\frac{1}{2}} \sinh x, \quad . \quad . \quad (6.10)$$

$$I_{-\frac{1}{2}}(x) = \left(\frac{2}{\pi x}\right)^{\frac{1}{2}} \cosh x. \quad . \quad . \quad (6.11)$$

§ 86. *Bessel's integral for* $J_n(x)$.

In accordance with the definition in § 83,

$$e^{\frac{x}{2}\left(t-\frac{1}{t}\right)} = \ldots + \frac{J_{-2}(x)}{t^2} + \frac{J_{-1}(x)}{t}$$
$$+ J_0(x) + J_1(x)t + J_2(x)t^2 + \ldots \quad (6.12)$$

which, by (6.4), may be written

$$e^{\frac{x}{2}\left(t-\frac{1}{t}\right)} = J_0(x) + J_1(x)\left(t - \frac{1}{t}\right) + J_2(x)\left(t^2 + \frac{1}{t^2}\right)$$
$$+ J_3(x)\left(t^3 - \frac{1}{t^3}\right) + \ldots \quad (6.13)$$

Making the substitution $t = e^{i\theta}$, we get

$$e^{ix\sin\theta} = J_0(x) + 2iJ_1(x)\sin\theta + 2J_2(x)\cos 2\theta + \ldots \quad (6.14)$$

and hence, by equating real and imaginary parts,

$$\cos(x\sin\theta) = J_0(x) + 2J_2(x)\cos 2\theta$$
$$+ 2J_4(x)\cos 4\theta + \ldots \quad (6.15)$$

$$\sin(x\sin\theta) = 2J_1(x)\sin\theta + 2J_3(x)\sin 3\theta + \ldots \quad (6.16)$$

These are Fourier expansions, and from them it follows, by the ordinary rule for finding the coefficients in a Fourier series, that, if n is even,

$$J_n(x) = \frac{1}{\pi}\int_0^\pi \cos(x\sin\theta)\cos n\theta\, d\theta, \quad . \quad (6.17)$$

$$0 = \frac{1}{\pi}\int_0^\pi \sin(x\sin\theta)\sin n\theta\, d\theta\,; \quad . \quad (6.18)$$

while, if n is odd,

$$0 = \frac{1}{\pi}\int_0^\pi \cos(x\sin\theta)\cos n\theta\, d\theta, \quad . \quad (6.19)$$

$$J_n(x) = \frac{1}{\pi}\int_0^\pi \sin(x\sin\theta)\sin n\theta\, d\theta. \quad . \quad (6.20)$$

By adding we find, if n is any positive integer,

$$J_n(x) = \frac{1}{\pi}\int_0^\pi \cos(n\theta - x\sin\theta)d\theta, \tag{6.21}$$

which is known as *Bessel's integral* for $J_n(x)$.

EXAMPLES X

1. By considering the expansion of the product

$$e^{\frac{x}{2}\left(t-\frac{1}{t}\right)} \cdot e^{\frac{y}{2}\left(t-\frac{1}{t}\right)}$$

in ascending and descending powers of t, in two ways, obtain the Addition formula for $J_0(x)$, viz.

(i) $J_0(x+y) = J_0(x)J_0(y) - 2J_1(x)J_1(y) + 2J_2(x)J_2(y) - \ldots$

By putting $y = -x$, $y = x$ in turn, show that

(ii) $1 = J_0^2(x) + 2J_1^2(x) + 2J_2^2(x) + \ldots$.

(iii) $J_0(2x) = J_0^2(x) - 2J_1^2(x) + 2J_2^2(x) - \ldots$

Show also that

(iv) $J_1(2x) = 2J_0(x)J_1(x) - 2J_1(x)J_2(x) + 2J_2(x)J_3(x) - \ldots$

[Compare the trigonometric formulæ $1 = \cos^2 x + \sin^2 x$, $\cos 2x = \cos^2 x - \sin^2 x$, $\sin 2x = 2\sin x\cos x$.]

2. Show that

$$J_0(x+iy) = J_0(x)I_0(y) - 2J_2(x)I_2(y) + 2J_4(x)I_4(y) - \ldots$$
$$- i\{2J_1(x)I_1(y) - 2J_3(x)I_3(y) + \ldots\}$$

3. Show that

$$e^{ix\cos\theta} = J_0(x) + 2\sum_{n=1}^{\infty} i^n J_n(x)\cos n\theta.$$

By considering the integral

$$\int_0^{2\pi} e^{ix\cos\theta} \cdot e^{-iy\cos(\theta-\alpha)}d\theta$$

in two ways, obtain the generalised Addition formula for $J_0(x)$, viz. :—

$$J_0\{\sqrt{(x^2+y^2-2xy\cos\alpha)}\} = J_0(x)J_0(y) + 2\sum_{n=1}^{\infty} J_n(x)J_n(y)\cos n\alpha.$$

4. Show that, if n is an integer,

$$J_n(x+y) = \sum_{p=-\infty}^{\infty} J_p(x)J_{n-p}(y).$$

5. By expanding $\cos n\theta$, $\sin n\theta$ on the right-hand side of (6.14) in ascending powers of $\sin \theta$, and then arranging both sides in ascending powers of $\sin \theta$, and equating coefficients, show that

$$\begin{aligned} 1 &= J_0(x) + 2J_2(x) + 2J_4(x) + 2J_6(x) + \dots \\ x &= 2\{J_1(x) + 3J_3(x) + 5J_5(x) + 7J_7(x) + \dots\} \\ x^2 &= 8\{J_2(x) + 4J_4(x) + 9J_6(x) + 16J_8(x) + \dots\} \\ &\dots\dots \end{aligned}$$

6. Obtain the expansions in the last example also by differentiating (6.14) with respect to θ repeatedly and putting $\theta = 0$.

7. By expanding $J_n(bx)$ and integrating term by term, show that, if $n > -1$, $a > 0$,

$$\int_0^\infty e^{-ax^2} J_n(bx) x^{n+1} dx = \frac{b^n}{(2a)^{n+1}} e^{-\frac{b^2}{4a}}.$$

8. Show from Ex. 1, (ii), that, if x is real and n a positive integer

$$|J_0(x)| < 1, \quad |J_n(x)| < 1/\sqrt{2}.$$

9. Prove from the series for $J_n(z)$ that, if $n > 0$, and $r = |z|$,

$$|J_n(z)| < \frac{r^n}{2^n \Gamma(n+1)} e^{\frac{1}{4}r^2}.$$

Deduce that $J_n(z) \to 0$ when $n \to +\infty$, for all values of z.

10. If $z = x + iy$, $(y > 0)$, and if $0 \leqslant t \leqslant 1$, prove that

$$|\cos zt| < e^y.$$

Hence, using Hankel's integral (§ 93) for $J_n(z)$, prove that

$$|J_n(z)| < \frac{r^n e^y}{2^n \Gamma(n+1)},$$

where $r = |z|$.

§ 87. *Bessel's differential equation of order n.*

If we write

$$y = x^n - \frac{x^{n+2}}{2 \,.\, 2n+2} + \frac{x^{n+4}}{2 \,.\, 4 \,.\, 2n+2 \,.\, 2n+4} - \dots \qquad (6.22)$$

we find at once, on differentiating term by term, that

$$x\frac{d}{dx}\left(x\frac{dy}{dx}\right) = n^2x^n - \frac{(n+2)^2x^{n+2}}{2 \,.\, 2n+2} + \frac{(n+4)^2x^{n+4}}{2 \,.\, 4 \,.\, 2n+2 \,.\, 2n+4} - \dots$$

Also

$$n^2y = n^2x^n - \frac{n^2x^{n+2}}{2\,.\,2n+2} + \frac{n^2x^{n+4}}{2\,.\,4\,.\,2n+2\,.\,2n+4} - \cdots$$

By subtraction,

$$x\frac{d}{dx}\left(x\frac{dy}{dx}\right) - n^2y = -\,x^{n+2} + \frac{x^{n+4}}{2\,.\,2n+2} - \cdots$$
$$= -\,x^2y,$$

and hence

$$x\frac{d}{dx}\left(x\frac{dy}{dx}\right) + (x^2 - n^2)y = 0, \quad . \quad . \quad (6{\cdot}23)$$

or

$$x^2\frac{d^2y}{dx^2} + x\frac{dy}{dx} + (x^2 - n^2)y = 0, \quad . \quad (6{\cdot}24)$$

which is known as *Bessel's equation of order n.*

Since $y = 2^n\Gamma(n+1)J_n(x)$, it follows that

$$x^2\frac{d^2J_n(x)}{dx^2} + x\frac{dJ_n(x)}{dx} + (x^2 - n^2)J_n(x) = 0\,; \quad (6.25)$$

in other words, $y = J_n(x)$ is one solution of Bessel's equation of order n. The general solution of this equation will be referred to later.

§ 88. Bessel's equation may also be written

$$\frac{d^2y}{dx^2} + \frac{1}{x}\frac{dy}{dx} + \left(1 - \frac{n^2}{x^2}\right)y = 0. \quad . \quad . \quad (6.26)$$

Replacing x by kx, we find that $y = J_n(kx)$ is one solution of the equation

$$\frac{d^2y}{dx^2} + \frac{1}{x}\frac{dy}{dx} + \left(k^2 - \frac{n^2}{x^2}\right)y = 0. \quad . \quad . \quad (6.27)$$

§ 89. *Recurrence formulæ.*

If we write down the series for $J_{n-1}(x)$ and $J_{n+1}(x)$, add and subtract them, and compare the sum with the series for $J_n(x)$, and the difference with the series for the derivative $J_n'(x)$, we find the following two formulæ, which, together with other forms in which they can be expressed, are known as the *recurrence formulæ* of the Bessel functions :—

$$\frac{2n}{x}J_n(x) = J_{n-1}(x) + J_{n+1}(x), \quad . \quad . \quad (6.28)$$

$$2J_n'(x) = J_{n-1}(x) - J_{n+1}(x). \quad . \quad . \quad (6.29)$$

By adding and subtracting and dividing by 2, we obtain the two formulæ

$$J_{n-1}(x) = \frac{n}{x}J_n(x) + J_n'(x), \quad . \quad . \quad (6.30)$$

$$J_{n+1}(x) = \frac{n}{x}J_n(x) - J_n'(x), \quad . \quad . \quad (6.31)$$

which express $J_{n-1}(x)$, $J_{n+1}(x)$ in terms of $J_n(x)$, $J_n'(x)$.

The last two formulæ can also be written

$$\frac{d}{dx}\{x^nJ_n(x)\} = x^nJ_{n-1}(x), \quad . \quad . \quad (6.32)$$

$$\frac{d}{dx}\left\{\frac{J_n(x)}{x^n}\right\} = -\frac{J_{n+1}(x)}{x^n}. \quad . \quad . \quad (6.33)$$

In particular,

$$\frac{d}{dx}\{xJ_1(x)\} = xJ_0(x), \quad . \quad . \quad (6.34)$$

$$\frac{d}{dx}J_0(x) = -J_1(x), \quad . \quad . \quad (6.35)$$

results which have already been noticed in § 3.

Inversely, we have

$$\int x^nJ_{n-1}(x)dx = x^nJ_n(x), \quad . \quad . \quad (6.36)$$

$$\int\frac{J_{n+1}(x)}{x^n}dx = -\frac{J_n(x)}{x^n}; \quad . \quad . \quad (6.37)$$

and hence also, if α is a constant $(\neq 0)$,

$$\int x^nJ_{n-1}(\alpha x)dx = \frac{x^nJ_n(\alpha x)}{\alpha}, \quad . \quad . \quad (6.38)$$

$$\int\frac{J_{n+1}(\alpha x)}{x^n}dx = -\frac{J_n(\alpha x)}{\alpha x^n}. \quad . \quad . \quad (6.39)$$

§ 90. By differentiating (6.31) repeatedly, and after each differentiation substituting for $J_n''(x)$ from the differential equation (6.25), it follows that $J_{n+1}(x)$ and all its derivatives can be expressed in the form

$$PJ_n(x) + QJ_n'(x),$$

where P, Q are polynomials in $1/x$.

Moreover, $J_{n-1}(x)$ and all its derivatives can be expressed in this form also, as we see by treating formula (6.30) in the same way.

COROLLARY 1. $J_n(x)$ and all its derivatives can be expressed in the form

$$PJ_p(x) + QJ_p'(x),$$

where P, Q are polynomials in $1/x$, and $n - p$ is any integer, positive or negative, or zero.

COROLLARY 2. If n is an integer, $J_n(x)$ and all its derivatives can be expressed in the form

$$PJ_0(x) + QJ_0'(x),$$

where P, Q are polynomials in $1/x$.

EXAMPLES XI

1. Show that

$$\text{(i)}\ J_0''(x) = -J_0(x) - \frac{1}{x}J_0'(x),$$

$$\text{(ii)}\ J_0'''(x) = \frac{1}{x}J_0(x) + \left(\frac{2}{x^2} - 1\right)J_0'(x).$$

2. Show that

$$\text{(i)}\ J_2(x) = \frac{2}{x}J_1(x) - J_0(x),$$

$$\text{(ii)}\ J_3(x) = \left(\frac{8}{x^2} - 1\right)J_1(x) - \frac{4}{x}J_0(x),$$

$$\text{(iii)}\ J_4(x) = \left(\frac{48}{x^3} - \frac{8}{x}\right)J_1(x) - \left(\frac{24}{x^2} - 1\right)J_0(x).$$

3. Show that

$$\text{(i)}\ J_2'(x) = \frac{2}{x}J_0(x) + \left(\frac{4}{x^2} - 1\right)J_0'(x),$$

$$\text{(ii)}\ J_3'(x) = \left(\frac{12}{x^2} - 1\right)J_0(x) + \left(\frac{24}{x^3} - \frac{5}{x}\right)J_0'(x).$$

4. Show that

(i) $J_2(x) = J_0(x) + 2J_0''(x)$,

(ii) $J_3(x) = 3J_1(x) + 4J_1''(x)$,

(iii) $J_{n+2}(x) = \left\{2n + 1 - \dfrac{2n(n^2 - 1)}{x^2}\right\}J_n(x) + 2(n + 1)J_n''(x)$.

5. Show that

$$J_{\frac{3}{2}}(x) = \left(\frac{2}{\pi x}\right)^{\frac{1}{2}}\left(\frac{\sin x}{x} - \cos x\right),$$

$$J_{-\frac{3}{2}}(x) = \left(\frac{2}{\pi x}\right)^{\frac{1}{2}}\left(- \sin x - \frac{\cos x}{x}\right),$$

and express $J_{\frac{5}{2}}(x)$, $J_{-\frac{5}{2}}(x)$ similarly.

6. Show that

$$\text{(i)} \quad \frac{J_{n+1}(x)}{x^{n+1}} = -\frac{1}{x}\frac{d}{dx}\frac{J_n(x)}{x^n}.$$

Deduce that

$$\text{(ii)} \quad \frac{J_{n+r}(x)}{x^{n+r}} = (-)^r\left(\frac{1}{x}\frac{d}{dx}\right)^r\frac{J_n(x)}{x^n},$$

$$\text{(iii)} \quad \frac{J_r(x)}{x^r} = (-)^r\left(\frac{1}{x}\frac{d}{dx}\right)^r J_0(x),$$

$$\text{(iv)} \quad J_{r+\frac{1}{2}}(x) = (-)^r\left(\frac{2}{\pi}\right)^{\frac{1}{2}}x^{r+\frac{1}{2}}\left(\frac{1}{x}\frac{d}{dx}\right)^r\left(\frac{\sin x}{x}\right).$$

7. Show that

(i) $2J_n'(x) = J_{n-1}(x) - J_{n+1}(x)$,

(ii) $4J_n''(x) = J_{n-2}(x) - 2J_n(x) + J_{n+2}(x)$,

(iii) $8J_n'''(x) = J_{n-3}(x) - 3J_{n-1}(x) + 3J_{n+1}(x) - J_{n+3}(x)$,

. . . . . .

8. Show that

(i) $xJ_0(x) = 2\{J_1(x) - 3J_3(x) + 5J_5(x) - \ldots\}$,

(ii) $\int J_0(x)dx = 2\{J_1(x) + J_3(x) + J_5(x) + \ldots\}$,

(iii) $\int xJ_0^2(x)dx = 2\{J_1^2(x) + 3J_3^2(x) + 5J_5^2(x) + \ldots\}$

Obtain corresponding results when $J_0(x)$ is replaced by $J_n(x)$ on the left-hand sides.

9. Show that

$$J_0(x)J_1(x) - J_n(x)J_{n+1}(x) = 2\sum_{r=1}^{n} J_r(x)J_r'(x).$$

Deduce that

$$\int_x^\infty J_n(t)J_{n+1}(t)dt = \tfrac{1}{2}J_0^2(x) + J_1^2(x) + \ldots + J_n^2(x),$$

$$n\int_x^\infty \frac{J_n^2(t)}{t}dt = \tfrac{1}{2}J_0^2(x) + J_1^2(x) + \ldots + J^2_{n-1}(x) + \tfrac{1}{2}J_n^2(x).$$

10. Show that

$$\int x^n J_n(x)dx = (2n - 1)\int x^{n-1} J_{n-1}(x)dx - x^n J_{n-1}(x).$$

Deduce that, if n is a positive integer,

$$\int x^n J_n(x)dx \text{ depends upon } \int J_0(x)dx.$$

Evaluate

$$\int_0^x t^n J_n(t)dt - \frac{(2n-1)!}{2^{n-1}(n-1)!}\int_0^x J_0(t)dt.$$

11. Show that if $n > 0$

$$\int_0^\infty J_{n+1}(x)dx = \int_0^\infty J_{n-1}(x)dx.$$

Deduce that, if n is a positive integer,

$$\int_0^\infty J_n(x)dx = 1$$

[for $n = 0$, see (4.7)] ; and also that

$$\int_0^\infty \frac{J_n(x)}{x}dx = \frac{1}{n}.$$

Hence, if $b > 0$, show that

$$\int_0^\infty J_n(bx)dx = \frac{1}{b}, \quad \int_0^\infty \frac{J_n(bx)}{x}dx = \frac{1}{n}.$$

12. Show that

$$\frac{2n}{x}I_n(x) = I_{n-1}(x) - I_{n+1}(x),$$

$$2I'_n(x) = I_{n-1}(x) + I_{n+1}(x).$$

13. Show that

$$I_{\frac{3}{2}}(x) = \left(\frac{2}{\pi x}\right)^{\frac{1}{2}}\left(\cosh x - \frac{\sinh x}{x}\right),$$

$$I_{-\frac{3}{2}}(x) = \left(\frac{2}{\pi x}\right)^{\frac{1}{2}}\left(\sinh x - \frac{\cosh x}{x}\right).$$

14. Show that $y = I_n(kx)$ is one solution of the equation

$$\frac{d^2y}{dx^2} + \frac{1}{x}\frac{dy}{dx} - \left(k^2 + \frac{n^2}{x^2}\right)y = 0.$$

(When $k = 1$, this equation is called the *modified Bessel equation* of order n.)

15. If n is an integer, show that $J_n(x)$ can be expressed in the form

$$J_n(x) = J_2(x)\Delta_{n-2} + J_1(x)\Delta_{n-3}$$

where Δ_{n-2}, Δ_{n-3} are polynomials in $1/x$; express them as determinants of order $n - 2$, $n - 3$ respectively.

§ 91. *Sonine's integral.*

Consider the integral

$$J_{n,p} = \int_0^1 (1 - x^2)^p \,.\, x^{n+1} J_n(\alpha x) dx, \qquad (6.40)$$

where $n > -1$ and α is a constant.

When $p = 0$, we find at once, using (6.38),

$$J_{n,0} = \frac{J_{n+1}(\alpha)}{\alpha}. \qquad (6.41)$$

Next, suppose that p is a positive integer. Then, by (6.38), we have

$$J_{n,p} = \int_0^1 (1 - x^2)^p \,.\, d\left\{\frac{x^{n+1}}{\alpha} J_{n+1}(\alpha x)\right\},$$

and hence, after integrating by parts,

$$J_{n,p} = \frac{2p}{\alpha} J_{n+1,\, p-1}$$

Applying this reduction formula repeatedly, we get

$$J_{n,p} = \frac{2^p p\,!}{\alpha^p} J_{n+p,\, 0}$$

that is, by (6.40) and (6.41),

$$\int_0^1 (1 - x^2)^p \,.\, x^{n+1} J_n(\alpha x) dx = \frac{2^p \Gamma(p + 1)}{\alpha^{p+1}} J_{n+p+1}(\alpha). \qquad (6.42)$$

Changing the notation by putting x for α, and t for x, we can write this result in the form

$$J_{n+p+1}(x) = \frac{x^{p+1}}{2^p \Gamma(p + 1)} \int_0^1 J_n(xt) \,.\, t^{n+1}(1 - t^2)^p dt. \qquad (6.43)$$

It holds good if $n > -1$, $p > -1$. For, although we have so far supposed p to be a positive integer, it is easy to verify that this restriction is unnecessary, by expanding $J_n(xt)$ and integrating term by term, using § 64. The conditions $n > -1$, $p > -1$, are necessary to ensure convergence at the lower and upper limits respectively.

Putting $m = n + p + 1$, we obtain another form of the last result, viz.

$$J_m(x) = \frac{2x^{m-n}}{2^{m-n}\Gamma(m-n)}\int_0^1 J_n(xt) \, . \, t^{n+1}(1-t^2)^{m-n-1}dt \quad (6.44)$$

where $m > n > -1$.

§ 92. The integral on the left of (6.42) or on the right of (6.43), (6.44) will be called here *Sonine's integral.** It is one of several integrals with which his name is associated in the theory of Bessel functions.

Sonine's integral enables us to express a Bessel function $J_m(x)$ as an integral involving $J_n(xt)$, of lower order, provided that $m > n > -1$.

§ 93. *Deductions from Sonine's integral.*

Put $n = -\frac{1}{2}$ in (6.44) ; then, since

$$J_{-\frac{1}{2}}(xt) = \left(\frac{2}{\pi xt}\right)^{\frac{1}{2}} \cos xt,$$

by (6.7), we find, if $m > -\frac{1}{2}$,

$$J_m(x) = \frac{x^m}{2^{m-1}\sqrt{\pi}\Gamma(m+\frac{1}{2})}\int_0^1 \cos xt \, . \, (1-t^2)^{m-\frac{1}{2}}dt, \quad (6.45)$$

which may be called *Hankel's integral* for $J_m(x)$.

Putting $t = \sin\theta$, we have, further, if $m > -\frac{1}{2}$,

$$J_m(x) = \frac{x^m}{2^{m-1}\sqrt{\pi}\Gamma(m+\frac{1}{2})}\int_0^{\frac{\pi}{2}} \cos(x\sin\theta)\cos^{2m}\theta \, d\theta, \quad (6.46)$$

which is known † as *Poisson's integral* for $J_m(x)$, or *Bessel's second integral* for $J_m(x)$.

* "Sonine's First Finite Integral," Watson, p. 373.

† Watson, p. 24.

EXAMPLES XII

1. Show that, if $n > -1$,

$$\text{(i)} \int_0^1 x^{n+1}J_n(\alpha x)dx = \frac{J_{n+1}(\alpha)}{\alpha},$$

$$\text{(ii)} \int_0^1 x^{n+1}(1 - x^2)J_n(\alpha x)dx = \frac{2J_{n+2}(\alpha)}{\alpha^2},$$

$$\text{(iii)} \int_0^1 x^{n+1}(1 - x^2)^2J_n(\alpha x)dx = \frac{8J_{n+3}(\alpha)}{\alpha^3}.$$

2. By writing

$$\begin{aligned} x^{2p} &= \{1 - (1 - x^2)\}^p \\ &= 1 - {}^pC_1(1 - x^2) + {}^pC_2(1 - x^2)^2 - \dots \end{aligned}$$

show that, if p is a positive integer,

$$2\int_0^1 x^{2p+1}J_0(\alpha x)dx$$

$$= \frac{2}{\alpha}J_1(\alpha) - p\left(\frac{2}{\alpha}\right)^2 J_2(\alpha) + p(p - 1)\left(\frac{2}{\alpha}\right)^3 J_3(\alpha) - \dots$$

In a similar way, when $n > -1$ find the value of the integral

$$\int_0^1 x^{n+2p+1}J_n(\alpha x)dx.$$

3. Show that, if $n > -1$,

$$\text{(i)} \int_0^1 x^{n+3}J_n(\alpha x)dx = \frac{J_{n+1}(\alpha)}{\alpha} - \frac{2J_{n+2}(\alpha)}{\alpha^2}.$$

$$\text{(ii)} \int_0^1 x^{n+5}J_n(\alpha x)dx = \frac{J_{n+1}(\alpha)}{\alpha} - \frac{4J_{n+2}(\alpha)}{\alpha^2} + \frac{8J_{n+3}(\alpha)}{\alpha^3}.$$

4. Show that

$$\int_0^a \frac{xJ_0(xy)}{\sqrt{(a^2 - x^2)}}dx = \frac{\sin ay}{y}.$$

5. Show that

$$\text{(i)} \int_0^{\frac{\pi}{2}} J_0(x \sin \theta) \sin \theta \, d\theta = \frac{\sin x}{x}.$$

$$\text{(ii)} \int_0^{\frac{\pi}{2}} J_1(x \sin \theta) \sin^2 \theta \, d\theta = \frac{\sin x - x \cos x}{x^2}.$$

$$\text{(iii)} \int_0^{\frac{\pi}{2}} J_n(x \sin \theta) \sin^{n+1}\theta \, d\theta = \left(\frac{\pi}{2x}\right)^{\frac{1}{2}} J_{n+\frac{1}{2}}(x).$$

6. Show that, if $n > 0$,

$$J_n(x) = \frac{x^n}{2^{n-1}\Gamma(n)}\int_0^1 J_0(xt) \,.\, t(1 - t^2)^{n-1}dt.$$

Deduce that there is a number ξ such that

$$J_n(x) = \frac{x^n J_0(\xi)}{2^n\Gamma(n + 1)}, \quad . \quad . \quad . \quad (6.47)$$

where $0 < \xi < x$.

7. By writing (6.44) in the form

$$J_m(x) = \frac{x^{m-n}}{2^{m-n}\Gamma(m - n)}\int_0^1 t^{-n}J_n(xt) \,.\, t^{2n}(1 - t^2)^{m-n-1}d(t^2),$$

and using the mean value theorem for integrals, show that there is a number ξ such that

$$\frac{2^m\Gamma(m + 1)}{x^m}J_m(x) = \frac{2^n\Gamma(n + 1)}{\xi^n}J_n(\xi), \quad . \quad . \quad (6.48)$$

where $0 < \xi < x$, $m > n > -1$.

8. By putting $m = \frac{1}{2}$, $x = \pi$ in (6.48), show that, if $-1 < n < \frac{1}{2}$, the equation $J_n(x) = 0$ has a root between 0 and π.

Also, by putting $m = -\frac{1}{2}$, $x = \frac{1}{2}\pi$, show that if $-1 < n < -\frac{1}{2}$, the equation $J_n(x) = 0$ has a root between 0 and $\frac{1}{2}\pi$.

9. If p is a positive integer, and $n > \frac{1}{2} + 2p$, show that

$$J_{n-p-1}(x) = \frac{x^{p+1}}{2^p\Gamma(p + 1)}\int_1^\infty \frac{J_n(xt)(t^2 - 1)^p}{t^{n-1}}dt.$$

Deduce that

$$\int_1^\infty J_n(xt)t^{-n+2p+1}dt$$
$$= \frac{1}{2}\left\{\frac{2}{x}J_{n-1}(x) + p\left(\frac{2}{x}\right)^2 J_{n-2}(x) + p(p - 1)\left(\frac{2}{x}\right)^3 J_{n-3}(x) + \ldots\right\}.$$

10. Assuming the first result in the last example to hold good * when $p > -1$, $n > \frac{1}{2} + 2p$, deduce that, if $-\frac{1}{2} < m < \frac{1}{2}$,

$$J_m(x) = \frac{2^{m+1}}{\sqrt{\pi}\,\Gamma(\frac{1}{2} - m)x^m}\int_1^\infty \frac{\sin xt\, dt}{(t^2 - 1)^{m+\frac{1}{2}}}.$$

* It can be deduced from Watson, p. 417, (5). The condition $p > -1$ is necessary for the convergence of the integral at the lower limit; $n > \frac{1}{2} + 2p$ is necessary for convergence at the upper limit, since $J_n(xt)$ behaves like $\dfrac{C\cos(xt - \lambda)}{\sqrt{(xt)}}$ when t is large (see § 95).

11. Making the same assumption as in the last example, show that

$$\int_a^\infty \frac{x\mathrm{J}_0(xy)dx}{\sqrt{(x^2 - a^2)}} = \frac{\cos ay}{y}.$$

§ 94. *Lommel's integrals.*

Put $u = \mathrm{J}_n(\alpha x)$, $v = \mathrm{J}_n(\beta x)$; then, by (6.27), using dashes to denote differentiation with respect to x, we can write the equations satisfied by u and v in the forms

$$xu'' + u' + \left(\alpha^2 - \frac{n^2}{x^2}\right)xu = 0, \qquad (6.49)$$

$$xv'' + v' + \left(\beta^2 - \frac{n^2}{x^2}\right)xv = 0, \qquad (6.50)$$

from which we find, by the method of § 11, that

$$(\beta^2 - \alpha^2)\int x\mathrm{J}_n(\alpha x)\mathrm{J}_n(\beta x)dx$$
$$= x\{\alpha\mathrm{J}_n'(\alpha x)\mathrm{J}_n(\beta x) - \beta\mathrm{J}_n'(\beta x)\mathrm{J}_n(\alpha x)\}. \quad (6.51)$$

Again, multiplying (6.49) throughout by $2xu'$, we have

$$2xu'\frac{d}{dx}(xu') + 2(\alpha^2x^2 - n^2)uu' = 0,$$

or

$$\frac{d}{dx}\{x^2u'^2 + (\alpha^2x^2 - n^2)u^2\} = 2\alpha^2xu^2.$$

Integrating, we get

$$x^2u'^2 + (\alpha^2x^2 - n^2)u^2 = 2\alpha^2\int xu^2dx,$$

and hence

$$\int x\mathrm{J}_n^2(\alpha x)dx = \frac{1}{2}\left\{x^2\mathrm{J}_n'^2(\alpha x) + \left(x^2 - \frac{n^2}{\alpha^2}\right)\mathrm{J}_n^2(\alpha x)\right\}, \quad (6.52)$$

which may also be written, with the aid of (6.30), (6.31),

$$\int x\mathrm{J}_n^2(\alpha x)dx = \frac{x^2}{2}\{\mathrm{J}_n^2(\alpha x) - \mathrm{J}_{n-1}(\alpha x)\mathrm{J}_{n+1}(\alpha x)\}. \quad (6.53)$$

In particular, when we integrate between the limits 0 and 1, we find from (6.51) and (6.53), provided $n > -1$ to ensure convergence at the lower limit,

$$(\beta^2 - \alpha^2)\int_0^1 x\mathrm{J}_n(\alpha x)\mathrm{J}_n(\beta x)dx$$
$$= \alpha\mathrm{J}_n'(\alpha)\mathrm{J}_n(\beta) - \beta\mathrm{J}_n'(\beta)\mathrm{J}_n(\alpha), \quad (6.54)$$
$$\int_0^1 x\mathrm{J}_n^2(\alpha x)dx = \tfrac{1}{2}\{\mathrm{J}_n^2(\alpha) - \mathrm{J}_{n-1}(\alpha)\mathrm{J}_{n+1}(\alpha)\}. \quad (6.55)$$

Examples XIII

1. If α, β, $(\beta^2 \neq \alpha^2)$, are two roots of the equation $\mathrm{J}_n(x) = 0$, show that, if $n > -1$,

$$\int_0^1 x\mathrm{J}_n(\alpha x)\mathrm{J}_n(\beta x)dx = 0.$$

Show also that the same result holds good if α, β, $(\beta^2 \neq \alpha^2)$, are two roots of the equation

$$x\mathrm{J}_n'(x) + \mathrm{H}\mathrm{J}_n(x) = 0,$$

where H is a constant.

2. Show that (6.54) can be written in the form

$$(\beta^2 - \alpha^2)\int_0^1 x\mathrm{J}_n(\alpha x)\mathrm{J}_n(\beta x)dx = \frac{\alpha\beta}{2n}\{\mathrm{J}_{n-1}(\alpha)\mathrm{J}_{n+1}(\beta) - \mathrm{J}_{n+1}(\alpha)\mathrm{J}_{n-1}(\beta)\}.$$

3. Derive (6.52) from (6.51) by differentiating partially with respect to β, and then putting $\beta = \alpha$.

4. Write down the differential equations satisfied by $\mathrm{J}_m(x)$, $\mathrm{J}_n(x)$, and show that, if $m + n > 0$,

$$(m^2 - n^2)\int_0^\alpha \frac{\mathrm{J}_m(x)\mathrm{J}_n(x)}{x}dx = \alpha\{\mathrm{J}_m'(\alpha)\mathrm{J}_n(\alpha) - \mathrm{J}_m(\alpha)\mathrm{J}_n'(\alpha)\}.$$

5. Show that

(i) $\displaystyle\int x\mathrm{J}_n^2(x)dx = \frac{x^2}{2}\{\mathrm{J}_n^2(x) - \mathrm{J}_{n-1}(x)\mathrm{J}_{n+1}(x)\}.$

(ii) $\displaystyle\int x^{2n+2}\mathrm{J}_n(x)\mathrm{J}_{n+1}(x)dx = \tfrac{1}{2}x^{2n+2}\mathrm{J}_{n+1}^2(x).$

(iii) $\displaystyle\int x^{2n+1}\mathrm{J}_n^2(x)dx = \frac{x^{2n+2}}{4n+2}\{\mathrm{J}_n^2(x) + \mathrm{J}_{n+1}^2(x)\}.$

(iv) $\displaystyle\int x^{2n+1}\mathrm{J}_{n-1}(x)\mathrm{J}_{n+1}(x)dx = \frac{x^{2n+2}}{4n+2}\{\mathrm{J}_{n-1}(x)\mathrm{J}_{n+1}(x) + \mathrm{J}_n(x)\mathrm{J}_{n+2}(x)\}.$

6. If $n > -1$, show that

$$\int_0^1 x\mathrm{I}_n^2(\alpha x)dx = \tfrac{1}{2}\{\mathrm{I}_n^2(\alpha) + \mathrm{I}_{n-1}(\alpha)\,\mathrm{I}_{n+1}(\alpha)\}.$$

7. Show that (cf. Exs. XI, 8)

$$\tfrac{1}{4}x^2\{\mathrm{J}_0^2(x) + \mathrm{J}_1^2(x)\} = \mathrm{J}_1^2(x) + 3\mathrm{J}_3^2(x) + 5\mathrm{J}_5^2(x) + \ldots$$

Show also that

$$\tfrac{1}{4}x^2\{J_n^2(x) - J_{n-1}(x)J_{n+1}(x)\}$$
$$= (n+1)J_{n+1}^2(x) + (n+3)J_{n+3}^2(x) + (n+5)J_{n+5}^2(x) + \dots$$

Deduce that

$$\frac{2x}{\pi} = J_{\frac{1}{2}}^2(x) + 3J_{\frac{3}{2}}^2(x) + 5J_{\frac{5}{2}}^2(x) + 7J_{\frac{7}{2}}^2(x) + \dots$$

$$\frac{\sin 2x}{\pi} = J_{\frac{1}{2}}^2(x) - 3J_{\frac{3}{2}}^2(x) + 5J_{\frac{5}{2}}^2(x) - 7J_{\frac{7}{2}}^2(x) + \dots$$

§ 95. *Large values of x.*

If we make the substitution

$$u = y\sqrt{x} \quad . \quad . \quad . \quad (6.56)$$

(cf. § 12) in Bessel's equation (6.24), we find that the equation satisfied by u is

$$\frac{d^2u}{dx^2} = -\left(1 - \frac{n^2 - \frac{1}{4}}{x^2}\right)u. \quad . \quad . \quad (6.57)$$

Now, when x is large enough, the term $(n^2 - \frac{1}{4})/x^2$ is as small as we please compared with 1, and then we have approximately

$$\frac{d^2u}{dx^2} = -u,$$

of which the general solution can be written $u = C\cos(x-\lambda)$, and we infer that every solution of Bessel's equation of order n can be written in the form

$$y = \frac{C}{\sqrt{x}}\{\cos(x-\lambda) + r(x)\}, \quad . \quad . \quad (6.58)$$

where C, λ are constants, and $r(x) \to 0$ when $x \to +\infty$. We deduce that

I. If $x > 0$, $y\sqrt{x}$ is bounded.

II. If $y = f(x)$ is any solution of Bessel's equation, the equation $f(x) = 0$ has an infinite number of real roots, and consecutive large roots differ by π approximately.

We shall now examine the roots of the particular equation $J_n(x) = 0$ more closely.

§ 96. *Roots of the equation* $J_n(x) = 0$, *when* n *is real.*

If α is a root of the equation $J_n(x) = 0$, it is plain from the series for $J_n(x)$ that $-\alpha$ is also a root. It will therefore suffice to confine our attention to positive roots.

I. *If* $k > 1$, *the equation* $J_n(kx) = 0$ *has at least one root between* a *and* $a + \pi$, *when* a *is sufficiently large.*

Proof. Put $u \equiv u(x) = \sqrt{(kx)}J_n(kx)$; then, replacing x by kx in (6.57), we find

$$\frac{d^2u}{dx^2} = -\left(k^2 - \frac{n^2 - \frac{1}{4}}{x^2}\right)u. \qquad . \qquad . \qquad (6{\cdot}59)$$

Also, put $v = \sin(x - a)$; then

$$\frac{d^2v}{dx^2} = -v. \qquad . \qquad . \qquad . \qquad (6.60)$$

Multiplying the first of these equations by v and the second by u and subtracting, we get

$$\frac{d}{dx}\left(u\frac{dv}{dx} - v\frac{du}{dx}\right) = \left(k^2 - 1 - \frac{n^2 - \frac{1}{4}}{x^2}\right)uv.$$

Now integrate between the limits $x = a$, $x = a + \pi$. Then, at the lower limit, $v = 0$, $dv/dx = 1$; at the upper limit $v = 0$, $dv/dx = -1$; hence

$$-u(a + \pi) - u(a) = \int_a^{a+\pi}\left(k^2 - 1 - \frac{n^2 - \frac{1}{4}}{x^2}\right)uv\,dx.$$

In the integrand on the right, u is continuous throughout the range of integration if $a > 0$; v is positive; and the rest of the integrand is positive if a is large enough. Consequently, by the mean value theorem for integrals, we have, for sufficiently large values of a,

$$-u(a + \pi) - u(a) = u(\xi)\int_a^{a+\pi}\left(k^2 - 1 - \frac{n^2 - \frac{1}{4}}{x^2}\right)vdx,$$

where $a < \xi < a + \pi$.

Since the integral on the right is now necessarily positive, it follows that $u(a)$, $u(\xi)$, $u(a + \pi)$ cannot all be of the same sign, and hence that the equation $u(x) = 0$ has at least one root between a and $a + \pi$. The theorem to be proved follows at once.

COROLLARY. *The equation* $J_n(x) = 0$ *has at least one root between* a *and* $a + k\pi$, *where* k *is any number greater than* 1, *and* a *is any sufficiently large number.*

II. *The equation* $J_n(x) = 0$ *has an infinite number of real roots, all simple.*

Proof. From I, Cor., it follows that, if $k > 1$, the equation $J_n(x) = 0$ has at least one root between every consecutive pair of terms of the sequence

$$a, \quad a + k\pi, \quad a + 2k\pi, \quad a + 3k\pi, \; . \; . \; .$$

provided a is sufficiently large. The equation $J_n(x) = 0$ has therefore an infinite number of positive real roots.

That all these roots are simple follows from the differential equation, as in § 14. (See also Exs. XIV, 3.)

III. *If* $n > -1$, *the equation* $J_n(x) = 0$ *has no purely imaginary root.*

Proof. This follows from the series for $J_n(x)$, by the same kind of reasoning as in § 15, II.

IV. *If* $n > -1$, *the equation* $J_n(x) = 0$ *has no complex roots.*

Proof. This follows from Exs. XIII, 1, by the same kind of reasoning as in § 15, III.

COROLLARY. From II, III, IV it follows that, if $n > -1$, the equation $J_n(x) = 0$ has an infinite number of simple real roots and no others, except possibly $x = 0$.

V. *If* n *is any real number, the equation* $J_n(x) = 0$ *has no root in common with either of the equations* $J_{n-1}(x) = 0$, $J_{n+1}(x) = 0$ (*except possibly* $x = 0$).

Proof. From the recurrence formulæ (6.30), (6.31), it would follow that a common root of $J_n(x) = 0$ and either of the equations $J_{n-1}(x) = 0$, $J_{n+1}(x) = 0$, would be also a root of $J_n'(x) = 0$, which is impossible, since all the roots of $J_n(x) = 0$ are simple (except possibly $x = 0$). (See also Exs. XIV, 3.)

VI. *If n is any real number, the equation $J_{n+1}(x) = 0$ has at least one root between every pair of positive roots of $J_n(x) = 0$.*

Proof. Since $J_n(x)/x^n$ and its derivative are continuous, it follows from the recurrence formula

$$\frac{d}{dx}\left\{\frac{J_n(x)}{x^n}\right\} = -\frac{J_{n+1}(x)}{x^n},$$

by Rolle's theorem, that the equation $J_{n+1}(x) = 0$ has at least one root between every pair of roots of $J_n(x) = 0$.

VII. *If n is any real number, the equation $J_{n-1}(x) = 0$ has at least one root between every pair of positive roots of $J_n(x) = 0$.*

Proof. This follows in the same kind of way from the recurrence formula

$$\frac{d}{dx}\{x^n J_n(x)\} = x^n J_{n-1}(x)$$

and Rolle's theorem.

VIII. *If n is any real number, the positive roots of $J_n(x) = 0$, $J_{n+1}(x) = 0$, interlace.*

Proof. From VI, the equation $J_{n+1}(x) = 0$ has at least one root between every adjacent pair of positive roots of $J_n(x) = 0$; and from VII, $J_n(x) = 0$ has at least one root between every adjacent pair of positive roots of $J_{n+1}(x) = 0$. It follows that between every adjacent pair of positive roots of either equation there lies one and only one root of the other, i.e. the roots of the two equations *interlace*.

IX. *If n is any real number, the equation*

$$xJ_n'(x) + HJ_n(x) = 0, \qquad . \qquad . \qquad (6.61)$$

where H is a real constant, has an infinite number of real roots.

This follows by the same kind of proof as in § 15, VII.

Note.—Equation (6.61) includes as particular cases

$$J_n'(x) = 0, \quad (H = 0);$$
$$J_{n-1}(x) = 0, \quad (H = n);$$
$$J_{n+1}(x) = 0, \quad (H = -n).$$

EXAMPLES XIV

1. If α is a root of $J_n(x) = 0$, show that

(i) $J_n'(\alpha) = J_{n-1}(\alpha) = -J_{n+1}(\alpha)$.

(ii) $\dfrac{J_{n+2}(\alpha)}{n+1} = \dfrac{2J_{n+1}(\alpha)}{\alpha} = -\dfrac{2J_{n-1}(\alpha)}{\alpha} = -\dfrac{J_{n-2}(\alpha)}{n-1}$.

2. If α is a root of $J_n'(x) = 0$, show that

$$\frac{nJ_n(\alpha)}{\alpha} = J_{n-1}(\alpha) = J_{n+1}(\alpha).$$

3. If α is a positive root of $J_n(x) = 0$, $(n > -1)$, show that

$$\int_0^1 xJ_n^2(\alpha x)dx = \tfrac{1}{2}J_n'^2(\alpha) = \tfrac{1}{2}J_{n+1}^2(\alpha) = \tfrac{1}{2}J_{n-1}^2(\alpha).$$

Deduce that the equation $J_n(x) = 0$ has no root in common with any of the equations $J_n'(x) = 0$, $J_{n+1}(x) = 0$, $J_{n-1}(x) = 0$, $xJ_n'(x) + HJ_n(x) = 0$, except possibly $x = 0$.

4. If α is a positive root of $J_n'(x) = 0$, $(n > -1)$, show that

$$\int_0^1 xJ_n^2(\alpha x)dx = \frac{\alpha^2 - n^2}{2\alpha^2}J_n^2(\alpha) = -\tfrac{1}{2}J_n(\alpha)J_n''(\alpha).$$

Deduce that, if α is the least positive root of $J_n'(x) = 0$, and β the least positive root of $J_n(x) = 0$, then $\beta > \alpha > n$.

Deduce also that the maximum values of $J_n(x)$ are all positive, and that the minimum values are all negative.

5. If α is a positive root of $J_n(x) = 0$, $(n > -1)$, show that

$$\int_0^1 xJ_n^2(\alpha x)dx = -\tfrac{1}{2}\alpha J_n'(\alpha)J_n''(\alpha).$$

Deduce that $J_n'(\alpha)$, $J_n''(\alpha)$ are of opposite sign, and interpret this on the graph of $J_n(x)$.

6. Show that the graphs of $J_{n+1}(x)$, $J_{n-1}(x)$ intersect at a point below each maximum point, and at a point above each minimum point on the graph of $J_n(x)$.

7. Show from (6.48) that the smallest positive root of the equation $J_n(x) = 0$ increases as n increases, if $n > -1$.

8. If α is a root of $J_n(x) = 0$, deduce from the result in Exs. XIII, 4, by differentiating partially with regard to m, and then putting $m = n$, that

$$\left[\frac{\partial J_n(x)}{\partial n}\right]_{x=\alpha} = -\frac{2n}{\alpha J_n'(\alpha)}\int_0^\alpha \frac{J_n^2(x)}{x}dx.$$

Hence show, by differentiating the equation $J_n(\alpha) = 0$, that

$$\frac{d\alpha}{dn} = \frac{2n}{\alpha J_{n+1}^2(\alpha)} \int_0^\alpha \frac{J_n^2(x)}{x} dx.$$

Deduce that, if $n > 0$, the positive roots of $J_n(x) = 0$ increase as n increases.

9. If $J_n(\alpha) = 0$, $(n > 0)$, show that

$$\int_0^1 xJ_{n-1}^2(\alpha x)dx = \int_0^1 xJ_n^2(\alpha x)dx = \int_0^1 xJ_{n+1}^2(\alpha x)dx.$$

10. If $y = f(x)$ is any solution of Bessel's equation of order n, show that the equation $f(x) = 0$ has one root between every consecutive pair of positive roots of the equation $J_n(x) = 0$. [See Exs. I, 7.]

§ 97. *Fourier-Bessel expansion of order n.*

Let $\alpha_1, \alpha_2, \alpha_3 \ldots$ denote the positive roots of the equation $J_n(x) = 0$, $(n > -1)$, arranged in ascending order of magnitude. Then it follows from (6.54) that, if $r \neq s$,

$$\int_0^1 xJ_n(x\alpha_r)J_n(x\alpha_s)dx = 0, \quad . \quad . \quad (6.62)$$

and from Exs. XIV, 3, that

$$\int_0^1 xJ_n^2(x\alpha_s)dx = \tfrac{1}{2}J_{n+1}^2(\alpha_s) \quad . \quad . \quad (6.63)$$

It can be proved (see § 99) that, if $n \geqslant -\frac{1}{2}$, a function $f(x)$ which is arbitrarily defined in the interval $0 < x < 1$, subject to certain conditions of integrability, can be expanded in an infinite series of the form

$$f(x) = A_1J_n(x\alpha_1) + A_2J_n(x\alpha_2) + A_3J_n(x\alpha_3) + \ldots \quad (6.64)$$

The coefficients A_s can be formally determined by multiplying throughout by $xJ_n(x\alpha_s)dx$ and integrating between the limits 0 and 1. For, we then have

$$\int_0^1 xf(x)J_n(x\alpha_s)dx = \sum_{r=1}^{\infty} A_r \int_0^1 xJ_n(x\alpha_r)J_n(x\alpha_s)dx.$$

By (6.62) every term on the right vanishes except the one in which $r = s$, and hence, by (6.63),

$$\int_0^1 xf(x)J_n(x\alpha_s)dx = \tfrac{1}{2}A_sJ_{n+1}^2(\alpha_s),$$

from which

$$A_s = \frac{2}{J_{n+1}^2(\alpha_s)}\int_0^1 xf(x)J_n(x\alpha_s)dx. \quad . \quad (6.65)$$

The expansion (6.64) is called the *Fourier-Bessel expansion* of $f(x)$ of order n.

§ 98. *Dini expansion of order n.*

An expansion similar to the Fourier-Bessel expansion, but based upon the roots of the equation

$$x^{-n}\{xJ_n'(x) + HJ_n(x)\} = 0, \quad . \quad . \quad (6.66)$$

is called the *Dini expansion* of $f(x)$ of order n.

Three cases may be distinguished, depending upon the values of the constant H :—

I. If $H > -n$, the Dini expansion has exactly the same form as the Fourier-Bessel expansion, viz.

$$f(x) = A_1J_n(x\alpha_1) + A_2J_n(x\alpha_2) + A_3J_n(x\alpha_3) + \ldots$$

where $\alpha_1, \alpha_2, \alpha_3, \ldots$ are the positive roots of (6.66).

By the same method as before, the determination of the constants A_s depends upon the integrals

$$\int_0^1 xJ_n(x\alpha_r)J_n(x\alpha_s)dx = 0, \quad (r \neq s),$$

(see Exs. XIII, 1), and

$$\int_0^1 xJ_n^2(x\alpha_s)dx = \tfrac{1}{2}\{J_n^2(\alpha_s) - J_{n-1}(\alpha_s)J_{n+1}(\alpha_s)\},$$

which follows from (6.55). Hence we find

$$A_s = \frac{2\int_0^1 xf(x)J_n(x\alpha_s)dx}{J_n^2(\alpha_s) - J_{n-1}(\alpha_s)J_{n+1}(\alpha_s)}. \quad (6.67)$$

II. If $H = -n$, equation (6.66) becomes

$$x^{-n}\{xJ_n'(x) - nJ_n(x)\} = 0,$$

that is,
$$x^{-n+1}J_{n+1}(x) = 0,$$

which has a double root $x = 0$, and in this case the Dini expansion has an initial term of the form $A_0 x^n$, so that now

$$f(x) = A_0 x^n + A_1 J_n(x\alpha_1) + A_2 J_n(x\alpha_2) + \ldots,$$

which may be regarded as an expansion based upon the positive roots (including zero) $0, \alpha_1, \alpha_2, \alpha_3, \ldots$ of the equation $J_{n+1}(x) = 0$.

The constant A_0 may be found by multiplying throughout by $x^{n+1}dx$ and integrating from 0 to 1; thus,

$$\int_0^1 f(x)x^{n+1}dx = A_0\int_0^1 x^{2n+1}dx = \frac{A_0}{2(n+1)},$$

which gives A_0, the remaining terms on the right vanishing in virtue of

$$\int_0^1 x^{n+1}J_n(x\alpha_s)dx = \frac{J_{n+1}(\alpha_s)}{\alpha_s} = 0, \quad (\alpha_s \neq 0).$$

The constants $A_s (s \neq 0)$ are now given by

$$\int_0^1 xf(x)J_n(x\alpha_s)dx = \tfrac{1}{2}A_s J_n{}^2(\alpha_s).$$

III. If $H < -n$, equation (6.66) has two purely imaginary roots, $\pm i\alpha_0$, and the Dini expansion begins with a term depending on them; * it is of the form

$$f(x) = A_0 I_n(x\alpha_0) + A_1 J_n(xx_1) + A_2 J_n(x\alpha_2) + \ldots$$

The coefficients A_s $(s \neq 0)$ are found as in I, and A_0 is found by multiplying throughout by $xI_n(x\alpha_0)dx$ and integrating from 0 to 1. [See Exs. XV, 8.]

§ 99. *Validity of the expansions.*

We shall not attempt here to establish the validity of the Fourier-Bessel and Dini expansions, but refer the reader to Watson, "Theory of Bessel Functions," Chap. XVIII, where the expansions are proved to be valid in the open interval $0 < x < 1$ when $n \geqslant -\frac{1}{2}$, provided that $f(x)$ has bounded variation in every closed interval contained in the open interval $0 < x < 1$, and that the integral $\int_0^1 | f(t) | \sqrt{t}\, dt$ exists.

It must be pointed out that these are not *necessary* conditions

* Watson, p. 597.

for the expansions to be valid: they are *sufficient* conditions on which a proof of the validity of the expansions can be based.

We may add that, if $f(x)$ is continuous, the Dini expansion converges uniformly to $f(x)$ in any interval $0 < a \leq x \leq 1$. The Fourier-Bessel expansion converges uniformly in any such interval if and only if the condition $f(1) = 0$ is satisfied, this condition being plainly necessary because the Fourier-Bessel series at $x = 1$ is $0 + 0 + 0 + \ldots$; if this condition is not satisfied, the interval of uniform convergence does not extend to the end point $x = 1$. For the end point $x = 0$, see Watson, § 18.55.

§ 100. *Example.*

One function that can be represented by a Fourier-Bessel or Dini expansion is $x^n(1 - x^2)^p$, $(n \geqslant -\frac{1}{2}, p > -1)$. For, by § 97 and (6.42), we find the Fourier-Bessel expansion

$$\frac{x^n(1 - x^2)^p}{\Gamma(p + 1)} = 2^{p+1}\sum_\alpha \frac{J_{n+p+1}(\alpha)J_n(\alpha x)}{\alpha^{p+1}J_{n+1}{}^2(\alpha)},$$

where the summation extends over the positive roots of $J_n(x) = 0$.

Again, by § 98, II, and (6.42), we find the Dini expansion

$$\frac{x^n(1 - x^2)^p}{\Gamma(p + 1)} = \frac{\Gamma(n + 2)x^n}{\Gamma(n + p + 2)} + 2^{p+1}\sum_\alpha \frac{J_{n+p+1}(\alpha)J_n(\alpha x)}{\alpha^{p+1}J_n{}^2(\alpha)},$$

where the summation extends over the positive roots of $J_{n+1}(x) = 0$.

EXAMPLES XV

1. Obtain the following (Fourier-Bessel) expansions, in which the summations extend over the positive roots of $J_n(x) = 0$.

(i) $x^n = 2\sum_\alpha \dfrac{J_n(\alpha x)}{\alpha J_{n+1}(\alpha)}$.

(ii) $x^n(1 - x^2) = 8(n + 1)\sum_\alpha \dfrac{J_n(\alpha x)}{\alpha^3 J_{n+1}(\alpha)}$.

(iii) $x^n(1 - x^2)^2 = 16\sum_\alpha \dfrac{J_{n+3}(\alpha)J_n(\alpha x)}{\alpha^3 J_{n+1}{}^2(\alpha)}$.

(iv) $x^{n+2p} = \sum_\alpha \dfrac{AJ_n(\alpha x)}{J_{n+1}{}^2(\alpha)}$, where

$$A = \frac{2}{\alpha}J_{n+1}(\alpha) - p\left(\frac{2}{\alpha}\right)^2 J_{n+2}(\alpha) + p(p - 1)\left(\frac{2}{\alpha}\right)^3 J_{n+3}(\alpha) - \ldots$$

and p is a positive integer [see Exs. XII, 2].

$$\text{(v)}\ J_n(kx) = 2J_n(k)\sum_\alpha \frac{\alpha J_n(\alpha x)}{(\alpha^2 - k^2)J_{n+1}(\alpha)}.$$

2. Show that, if $p > -1$,

$$(1 - x^2)^p = \frac{1}{p+1} + 2^{p+1}\Gamma(p+1)\sum_\alpha \frac{J_{p+1}(\alpha)J_0(\alpha x)}{\alpha^{p+1}J_0^2(\alpha)}$$

where the summation extends over the positive roots of $J_1(x) = 0$.

3. Show that, if $n \geqslant -\frac{1}{2}$,

$$x^n(1 - x^2)^2 = \frac{2x^n}{(n+2)(n+3)} - 32(n+2)\sum_\alpha \frac{J_n(\alpha x)}{\alpha^4 J_n(\alpha)}$$

where the summation extends over the positive roots of the equation $J_{n+1}(x) = 0$.

4. Show that, if $-1 < x < 1$,

$$\text{(i)}\ \frac{x}{\sqrt{(1 - x^2)}} = \pi\sum_{s=1}^{\infty} J_1(s\pi)\sin s\pi x;$$

$$\text{(ii)}\ \frac{x}{\sqrt{(1 - x^2)}} = \pi\sum_{s=1}^{\infty} J_1(r\pi)\sin r\pi x,$$

where $r = s - \frac{1}{2}$;

$$\text{(iii)}\ \frac{x}{\sqrt{(1 - x^2)}} = \pi\left\{\frac{3x}{4} + \sum_\alpha \frac{J_1(\alpha)\sin\alpha x}{\sin^2\alpha}\right\}$$

where the summation extends over the positive roots of the equation $x = \tan x$.

5. Show that, if $-1 < x < 1$,

$$\text{(i)}\ \frac{1}{\sqrt{(1 - x^2)}} = \pi\sum_{s=1}^{\infty} J_0(r\pi)\cos r\pi x$$

where $r = s - \frac{1}{2}$;

$$\text{(ii)}\ \frac{1}{\sqrt{(1 - x^2)}} = \frac{\pi}{2} + \pi\sum_{s=1}^{\infty} J_0(s\pi)\cos s\pi x.$$

6. If $q > 0$, $-1 < x < 1$, and $J_0(\alpha) = 0$, show that

$$(1 - x^2)^{q-1} = 2^q\Gamma(q)\sum_\alpha \frac{J_q(\alpha)J_0(\alpha x)}{\alpha^q J_1^2(\alpha)}.$$

Deduce that

$$\frac{1}{\sqrt{(1 - x^2)}} = 2\sum_\alpha \frac{\sin\alpha}{\alpha}\frac{J_0(\alpha x)}{J_1^2(\alpha)}.$$

Multiply throughout by $x dx$, and by integration deduce that

$$\frac{1 - \sqrt{(1 - x^2)}}{x} = 2\sum_{\alpha} \frac{\sin \alpha}{\alpha^2} \frac{J_1(\alpha x)}{J_1^2(\alpha)}.$$

7. Deduce the Fourier sine and cosine series as particular cases of the Fourier-Bessel expansion.

8. Assuming the validity of the Dini expansion in III, § 98, show that the coefficient A_0 is given by

$$A_0 = \frac{2\int_0^1 x f(x) I_n(x\alpha_0) dx}{I_n^2(\alpha_0) + I_{n-1}(\alpha_0) I_{n+1}(\alpha_0)}.$$

§ 101. *The Fourier-Bessel double integral.*

The following argument * shows the plausibility of expressing an arbitrary function in the form of a double integral analogous to the Fourier double integral.

Let a function $f(x)$ be defined from $x = 0$ to $x = h$, by

$$f(x) = \phi(x), \quad 0 \leqslant x < a, \quad . \quad . \quad \text{(i)}$$

$$f(x) = 0, \quad a \leqslant x \leqslant h. \quad . \quad . \quad \text{(ii)}$$

Then by Exs. II, 5, we have

$$f(x) = \sum_{\alpha} \frac{2}{h^2 J_1^2(\alpha)} \int_0^h f(t) J_0\left(\frac{x\alpha}{h}\right) J_0\left(\frac{t\alpha}{h}\right) t dt,$$

where the summation extends over the positive roots of the equation $J_0(x) = 0$, and $0 < x < h$. It follows from (ii) that, if $0 < x < a < h$,

$$\phi(x) = \sum_{\alpha} \frac{2}{h^2 J_1^2(\alpha)} \int_0^a \phi(t) J_0\left(\frac{x\alpha}{h}\right) J_0\left(\frac{t\alpha}{h}\right) t dt.$$

Now every term of this series tends to zero when $h \to \infty$. Consequently, when h is large, a finite number of terms at the beginning of the series can be neglected. Hence, if for α_s in the sth term we make the substitution (5.37)

$$\alpha_s = (s - \tfrac{1}{4})\pi,$$

the error that we make by this substitution in the first few terms of the series will vanish when $h \to \infty$.

* Cf. Riemann-Weber, I, p. 199 (6th edn.).

Again, by (5.36), with this substitution,

$$J_1^2(\alpha_s) \doteq \frac{2}{\pi\alpha_s}\sin^2\left(\alpha_s - \frac{\pi}{4}\right) = \frac{2}{\pi\alpha_s},$$

and hence

$$\phi(x) = \sum_{s=1}^{\infty} \frac{\pi\alpha_s}{h^2}\int_0^a \phi(t)J_0\left(\frac{x\alpha_s}{h}\right)J_0\left(\frac{t\alpha_s}{h}\right)tdt.$$

Now put

$$y_s = \frac{\alpha_s}{h}, \quad \delta y = y_{s+1} - y_s = \frac{\pi}{h},$$

then $$\phi(x) = \sum_{s=1}^{\infty} y_s\delta y\int_0^a \phi(t)J_0(xy_s)J_0(ty_s)tdt,$$

and we infer that, in the limit when $h \to \infty$,

$$\phi(x) = \int_0^{\infty} J_0(xy)ydy\int_0^a \phi(t)J_0(yt)tdt. \quad . \quad (6.68)$$

If the function $\phi(x)$ is suitable (see Note below), we can put $a = \infty$, and we then have a function, defined from 0 to ∞, expressed as a double integral, thus

$$\phi(x) = \int_0^{\infty} J_0(xy)ydy\int_0^{\infty} \phi(t)J_0(yt)tdt \quad . \quad (6.69)$$

This formula and the more general formula

$$\phi(x) = \int_0^{\infty} J_n(xy)ydy\int_0^{\infty} \phi(t)J_n(yt)tdt \quad . \quad (6.70)$$

can be proved rigorously under certain conditions.

Note.—Sufficient conditions on which a proof of the validity of (6.70), when $x > 0$, can be based are: $n > \frac{1}{2}$, $\phi(t)$ has bounded variation in every closed interval in which t is positive, and $\int_0^{\infty} | \phi(t) | \sqrt{t}\, dt$ exists. [See Watson, § 14.4.]

Examples XVI

1. Verify (6.70) when

$$\text{(i) } \phi(t) = \frac{1}{t}; \quad \text{(ii) } \phi(t) = t^n e^{-at^2}.$$

[Use Exs. XI, 11; Exs. X, 7.]

2. Show from (6.70) that if

$$\int_0^\infty \phi(x)J_n(xy)xdx = \psi(y),$$

then
$$\int_0^\infty \psi(x)J_n(xy)xdx = \phi(y).$$

(i) By putting $n = 0$, $\phi(x) = e^{-ax}/x$, verify that, if $a > 0$, $b > 0$,

$$\int_0^\infty \frac{xJ_0(bx)}{\sqrt{(a^2 + x^2)}}dx = \frac{e^{-ab}}{b}.$$

[See (4.5) and Exs. VII, 3.]

(ii) By putting $n = 0$, $\phi(x) = \dfrac{\sin ax}{x}$, verify that

$$\int_0^a \frac{xJ_0(xy)}{\sqrt{(a^2 - x^2)}}dx = \frac{\sin ay}{y}.$$

[See (4.21) and Exs. XII, 4.]

(iii) By putting $n = 0$, $\phi(x) = \dfrac{\cos ax}{x}$, verify that

$$\int_a^\infty \frac{xJ_0(xy)dx}{\sqrt{(x^2 - a^2)}} = \frac{\cos ay}{y}.$$

[See (4.19) and Exs. XII, 11.]

§ 102. *General solution of Bessel's equation.*

We have seen (§ 87) that $y = J_n(x)$ is a solution of Bessel's differential equation,

$$\frac{d^2y}{dx^2} + \frac{1}{x}\frac{dy}{dx} + \left(1 - \frac{n^2}{x^2}\right)y = 0. \quad . \quad . \quad (6.71)$$

Since this equation remains unaltered when n is replaced by $-n$, a second solution is $y = J_{-n}(x)$, and if n is not an integer the *general solution* can be written

$$y = AJ_n(x) + BJ_{-n}(x). \quad . \quad . \quad (6.72)$$

But, *if n is an integer*, $J_{-n}(x) = (-)^nJ_n(x)$, and (6.72) is no longer the general solution. In this case, by the same method as in § 4, the general solution can be written in the form

$$y = AJ_n(x) + BJ_n(x)\int\frac{dx}{xJ_n^2(x)}. \quad . \quad . \quad (6.73)$$

If we substitute for $J_n(x)$ its series in the last integral, and expand the integrand in ascending powers of x, the integral takes the form

$$\int\left(\frac{a_n}{x^{2n+1}} + \ldots + \frac{a_1}{x^3} + \frac{a_0}{x} + b_1x + b_2x^3 + b_3x^5 + \ldots\right)dx$$

$$= -\frac{a_n}{2nx^{2n}} - \ldots - \frac{a_1}{2x^2} + a_0 \log x$$

$$+ b_0 + \frac{b_1x^2}{2} + \frac{b_2x^4}{4} + \ldots$$

where b_0 is a constant of integration. When this expression is multiplied by $J_n(x)$ and the product substituted in the second term of (6.73), we obtain a second solution of the form

$$J_n(x)(a_0 \log x + b_0) + \frac{1}{x^n}(c_0 + c_1x^2 + c_2x^4 + \ldots)$$

Here a_0, c_0, c_1, c_2, . . . are definite constants; the constant b_0, however, is arbitrary, and its value can be chosen so as to give the most convenient form to the second solution. The form which is now generally accepted as the standard one is known as Weber's, and is denoted by $Y_n(x)$. In terms of $J_n(x)$ and $Y_n(x)$, the general solution takes the form

$$y = AJ_n(x) + BY_n(x). \quad . \quad . \quad (6.74)$$

There is no need here to determine the actual values of the coefficients in the expansion of $Y_n(x)$. It is sufficient to remember that any second solution of the equation, when n is an integer, behaves like $1/x^n$ when x is small, and involves $\log x$.

§ 103. If we replace x by kx in (6.71), we obtain the equation

$$\frac{d^2y}{dx^2} + \frac{1}{x}\frac{dy}{dx} + \left(k^2 - \frac{n^2}{x^2}\right)y = 0, \quad . \quad (6.75)$$

of which the general solution may be written

$$y = AJ_n(kx) + BY_n(kx), \quad . \quad . \quad (6.76)$$

or

$$y = AJ_n(kx) + BJ_{-n}(kx), \quad . \quad . \quad (6.77)$$

according as n is an integer or not.

§ 104. *Transformations of Bessel's equation.*

A number of transformations of Bessel's equation, together with their solutions, can be obtained by finding the equation satisfied by

$$y = x^{\alpha}\mathrm{J}_n(\beta x^{\gamma}) \quad . \quad . \quad . \quad (6.78)$$

where α, β, γ are constants. If we put

$$\eta = \frac{y}{x^{\alpha}}, \quad \xi = \beta x^{\gamma}, \quad . \quad . \quad . \quad (6.79)$$

(6.78) gives $\eta = \mathrm{J}_n(\xi)$, and hence

$$\xi^2\frac{d^2\eta}{d\xi^2} + \xi\frac{d\eta}{d\xi} + (\xi^2 - n^2)\eta = 0,$$

which may be written

$$\xi\frac{d}{d\xi}\left(\xi\frac{d\eta}{d\xi}\right) + (\xi^2 - n^2)\eta = 0.$$

Now
$$\xi\frac{d\eta}{d\xi} = \xi\frac{d\eta/dx}{d\xi/dx} = \frac{1}{\gamma}\cdot x\frac{d\eta}{dx},$$

by (6.79), and hence

$$\xi\frac{d}{d\xi}\left(\xi\frac{d\eta}{d\xi}\right) = \frac{1}{\gamma^2}x\frac{d}{dx}\left(x\frac{d\eta}{dx}\right).$$

Again, from (6.79) we find

$$x\frac{d\eta}{dx} = \frac{y'}{x^{\alpha-1}} - \frac{\alpha y}{x^{\alpha}},$$

and further,

$$x\frac{d}{dx}\left(x\frac{d\eta}{dx}\right) = \frac{y''}{x^{\alpha-2}} - \frac{(2\alpha - 1)y'}{x^{\alpha-1}} + \frac{\alpha^2 y}{x^{\alpha}}.$$

Hence the equation satisfied by y is

$$\frac{1}{\gamma^2}\left(\frac{y''}{x^{\alpha-2}} - \frac{(2\alpha - 1)y'}{x^{\alpha-1}} + \frac{\alpha^2 y}{x^{\alpha}}\right) + (\beta^2 x^{2\gamma} - n^2)\frac{y}{x^{\alpha}} = 0,$$

or

$$\frac{d^2y}{dx^2} - \frac{2\alpha - 1}{x}\frac{dy}{dx} + \left(\beta^2\gamma^2 x^{2\gamma-2} + \frac{\alpha^2 - n^2\gamma^2}{x^2}\right)y = 0. \quad (6.80)$$

The general solution of this equation is

$$y = x^{\alpha}\{AJ_n(\beta x^{\gamma}) + BY_n(\beta x^{\gamma})\}, \quad . \quad . \quad (6.81)$$

or

$$y = x^{\alpha}\{AJ_n(\beta x^{\gamma}) + BJ_{-n}(\beta x^{\gamma})\}, \quad . \quad (6.82)$$

by §103, according as n is an integer or not.

§ 105. Particular cases of equations whose solutions can at once be written down in terms of Bessel functions are obtained by giving particular values to the constants α, β, γ, n. For example,

(i) $\alpha = 0, \gamma = 1$, gives

$$\frac{d^2y}{dx^2} + \frac{1}{x}\frac{dy}{dx} + \left(\beta^2 - \frac{n^2}{x^2}\right)y = 0;$$

(ii) $\alpha = \frac{1}{2}$ gives

$$\frac{d^2y}{dx^2} + \left(\beta^2\gamma^2x^{2\gamma-2} + \frac{\frac{1}{4} - n^2\gamma^2}{x^2}\right)y = 0;$$

(iii) $\alpha = \frac{1}{2},\ \beta = 1,\ \gamma = 1$, gives

$$\frac{d^2y}{dx^2} + \left(1 - \frac{n^2 - \frac{1}{4}}{x^2}\right)y = 0;$$

(iv) $\alpha = \frac{1}{2},\ \beta = \frac{2k}{m+2},\ \gamma = \frac{m+2}{2},\ n = \frac{1}{m+2}$, gives

$$\frac{d^2y}{dx^2} + k^2x^my = 0;$$

(v) $\alpha = n,\ \beta = 1,\ \gamma = 1$, gives

$$\frac{d^2y}{dx^2} - \frac{2n-1}{x}\frac{dy}{dx} + y = 0;$$

(vi) $\alpha = -n, \beta = 1, \gamma = 1$, gives

$$\frac{d^2y}{dx^2} + \frac{2n+1}{x}\frac{dy}{dx} + y = 0.$$

Examples XVII

1. Show that the general solution of the equation

$$4\frac{d^2y}{dx^2} + 9xy = 0$$

can be written

$$y = \sqrt{x}\{AJ_{-\frac{1}{3}}(x^{\frac{3}{2}}) + BJ_{\frac{1}{3}}(x^{\frac{3}{2}})\}.$$

Also solve the equation in series.

2. Show that the general solution of the equation

$$\frac{d^2y}{dx^2} + 4x^2y = 0$$

is

$$y = \sqrt{x}\{\mathrm{AJ}_{-\frac{1}{4}}(x^2) + \mathrm{BJ}_{\frac{1}{4}}(x^2)\}.$$

Also solve the equation in series.

3. If y satisfies the equation

$$\frac{d^2y}{dx^2} + x^4y = ax^5$$

and if $dy/dx = a$ when $x = 0$, show that y can be written in either of the forms

$$y = ax + \mathrm{B}\left(1 - \frac{x^6}{5 . 6} + \frac{x^{12}}{5 . 6 . 11 . 12} - \dots\right),$$

$$y = ax + \mathrm{C}\sqrt{x}\mathrm{J}_{-\frac{1}{6}}\left(\frac{x^3}{3}\right),$$

where B, or C, is an arbitrary constant.

4. Show that Riccati's equation

$$\frac{dy}{dx} + by^2 = cx^m$$

is transformed into

$$\frac{d^2u}{dx^2} - bcx^mu = 0$$

by the substitution

$$by = \frac{1}{u}\frac{du}{dx}.$$

Hence show how the solution of Riccati's equation can be expressed in terms of Bessel functions.

5. Use the last example to show that the general solution of the equation

$$\frac{dy}{dx} = x^2 + y^2$$

can be written

$$y = x . \frac{\mathrm{AJ}_{-\frac{3}{4}}(\frac{1}{2}x^2) + \mathrm{J}_{\frac{3}{4}}(\frac{1}{2}x^2)}{\mathrm{J}_{-\frac{1}{4}}(\frac{1}{2}x^2) - \mathrm{AJ}_{\frac{1}{4}}(\frac{1}{2}x^2)}$$

where A is an arbitrary constant.

From this solution, or by solving the equation in series, verify that, if $y = a$ when $x = 0$,

$$y = a + a^2x + a^3x^2 + (a^4 + \tfrac{1}{3})x^3 + \dots$$

If $y = 0$ when $x = 0$, show that

$$y = \frac{x^3}{3} + \frac{x^7}{63} + \frac{2x^{11}}{2079} + \dots$$

§ 106. *Contour integral for* $J_n(x)$.

It follows from the definition of $J_n(x)$ in § 83 that, when n is an integer, $J_n(x)$ is the residue of the function

$$\frac{e^{\frac{x}{2}\left(t-\frac{1}{t}\right)}}{t^{n+1}} \quad . \quad . \quad . \quad . \quad (6.83)$$

at $t = 0$, and hence that $J_n(x)$ can be expressed as a contour integral in the form

$$J_n(x) = \frac{1}{2\pi i}\int_C e^{\frac{x}{2}\left(t-\frac{1}{t}\right)}\frac{dt}{t^{n+1}} \quad . \quad . \quad (6.84)$$

where C denotes any simple contour surrounding the origin.

Moreover, by a simple modification, $J_n(x)$ can be expressed as a contour integral which can be regarded as *defining* $J_n(x)$ for all values of n, real or complex.* It is, however, beyond the scope of this book to pursue the study of Bessel functions further from this point of view.

§ 107. *Hankel's contour integral.*

From (6.45) it follows that

$$J_n(x) = \frac{x^n}{2^n\sqrt{\pi}\Gamma(n+\frac{1}{2})}\int_{-1}^{1} e^{ixt}(1-t^2)^{n-\frac{1}{2}}dt. \quad (6.85)$$

We can verify that this expression satisfies Bessel's differential equation; for if we put

$$y = x^n\int_{-1}^{1} e^{ixt}(1-t^2)^{n-\frac{1}{2}}dt,$$

we find by differentiation

$$x^2\frac{d^2y}{dx^2} + x\frac{dy}{dx} + (x^2 - n^2)y$$

$$= -ix^{n+1}\int_{-1}^{1}\{ixe^{ixt}(1-t^2)^{n+\frac{1}{2}} - (2n+1)te^{ixt}(1-t^2)^{n-\frac{1}{2}}\}dt$$

$$= -ix^{n+1}\int_{-1}^{1}\frac{d}{dt}\{e^{ixt}(1-t^2)^{n+\frac{1}{2}}\}dt$$

$$= -ix^{n+1}\Big[e^{ixt}(1-t^2)^{n+\frac{1}{2}}\Big]_{-1}^{1}$$

$$= 0, \text{ if } n > -\tfrac{1}{2}.$$

* Whittaker and Watson: "Modern Analysis," § 17.2.

§ 108. More generally, if we put

$$y = x^n \int_a^b e^{ixt}(1 - t^2)^{n-\frac{1}{2}}dt,$$

we find that

$$x^2\frac{d^2y}{dx^2} + x\frac{dy}{dx} + (x^2 - n^2)y = -ix^{n+1}\Big[e^{ixt}(1 - t^2)^{n+\frac{1}{2}}\Big]_a^b,$$

and hence, if $x > 0$, that y is a solution of Bessel's equation of order n if the integral is taken along any contour from either of the points $t = \pm 1$ to an infinitely distant point in the upper half of the t-plane (cf. § 69).

Accordingly, we can define further solutions (Hankel functions) when $n > -\frac{1}{2}$, $x > 0$, by the formulæ

$$-\mathrm{H}_n^{(1)}(x) = \frac{x^n}{2^{n-1}\sqrt{\pi}\Gamma(n + \frac{1}{2})}\int_1^{1+i\infty} e^{ixt}(1 - t^2)^{n-\frac{1}{2}}dt,$$

$$-\mathrm{H}_n^{(2)}(x) = \frac{x^n}{2^{n-1}\sqrt{\pi}\Gamma(n + \frac{1}{2})}\int_{-1+i\infty}^{-1} e^{ixt}(1 - t^2)^{n-\frac{1}{2}}dt.$$

These functions of order n correspond to $\mathrm{H}_0^{(1)}(x)$, $\mathrm{H}_0^{(2)}(x)$ in the theory of Bessel functions of zero order (§ 73). From them we can, for example, develop the asymptotic expansions of order n.

CHAPTER VII

APPLICATIONS

§ 109. *Kepler's problem.*

In the ideal problem of planetary motion a planet P moves in an ellipse under the gravitational attraction of a sun S situated in one of the foci, the area swept out by the radius vector SP during any interval of time being proportional to that interval (Fig. 18).

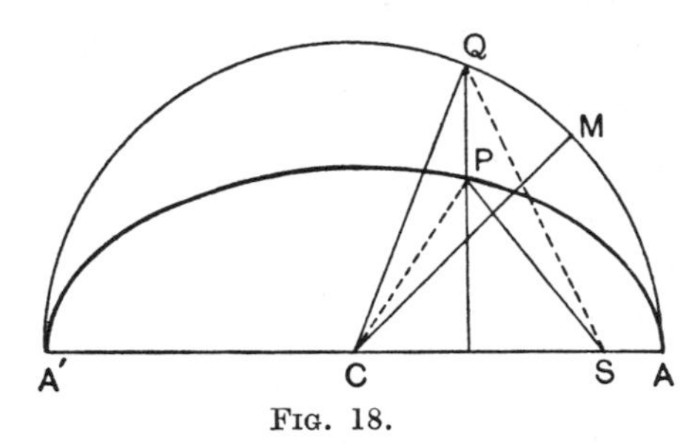

FIG. 18.

Let A′A be the major axis of the ellipse, and let a line drawn through P perpendicular to A′A meet the circle described on A′A as diameter in the point Q. Draw CP, CQ, SQ. In the usual notation of the ellipse, let $2a$ be the length of the major axis A′A, $2b$ that of the minor axis, and e the eccentricity.

Further, let a point M describe the circle AQA′ at such a constant speed that it coincides with P at A and A′. Let the time t be measured from an instant when P is passing through perihelion at A, and let $r = \text{SP}$, $\theta = \text{angle ASP}$, $\phi = \text{angle ACQ}$, $\psi = \text{angle ACM}$, all measured at time t.

In the terminology of astronomy, θ is called the *true anomaly*, ϕ the *eccentric anomaly*, and ψ, which is proportional to t, the *mean anomaly*.

Kepler's problem was to express such variables as r, θ, ϕ explicitly in terms of the time t, or, what comes to the same thing, in terms of the mean anomaly ψ.

§ 110. *Bessel's solution.*

Consider, for example, the expression of ϕ in terms of ψ.

Since ψ and the area ASP are both proportional to t, we have, if T is the period in which the complete ellipse is described,

$$\frac{t}{T} = \frac{\psi}{2\pi} = \frac{\text{area ASP}}{\pi ab} = \frac{\text{area ASQ}}{\pi a^2}.$$

Now $$\text{area ASQ} = \text{sector ACQ} - \triangle\text{SCQ}$$
$$= \tfrac{1}{2}a^2\phi - \tfrac{1}{2}ae \,.\, a \sin \phi,$$

and hence $$\psi = \phi - e \sin \phi. \quad . \quad . \quad . \quad (7.1)$$

The problem is now to solve this equation for ϕ, so as to express ϕ explicitly in terms of ψ.

It is evident that $\phi - \psi$ is an odd periodic function of ϕ, with period 2π, and may therefore be expanded in a Fourier sine-series. Consequently, we put

$$\phi - \psi = B_1 \sin \psi + B_2 \sin 2\psi + B_3 \sin 3\psi + \ldots \quad (7.2)$$

The coefficients B_n are now given, in accordance with the usual rule, by

$$\frac{\pi}{2}B_n = \int_0^\pi (\phi - \psi) \sin n\psi \, d\psi$$
$$= \left[-(\phi - \psi)\frac{\cos n\psi}{n}\right]_0^\pi + \int_0^\pi \frac{\cos n\psi}{n}(d\phi - d\psi).$$

In the first term on the right, $\phi - \psi$ vanishes at both limits, and in the second term the integral of $\cos n\psi \, d\psi$ between the limits 0 and π, is zero ; consequently

$$B_n = \frac{2}{\pi n}\int_0^\pi \cos n\psi \, d\phi$$
$$= \frac{2}{\pi n}\int_0^\pi \cos (n\phi - ne \sin \phi) d\phi,$$

by (7.1), and hence, by (6.21),

$$B_n = \frac{2}{n}J_n(ne). \quad . \quad . \quad . \quad (7.3)$$

Putting $n = 1, 2, 3, \ldots$ and substituting the values of $B_1, B_2, B_3, \ldots$ in (7.2), we obtain

$$\phi = \psi + 2\left\{J_1(e)\frac{\sin \psi}{1} + J_2(2e)\frac{\sin 2\psi}{2} + J_3(3e)\frac{\sin 3\psi}{3} + \ldots\right\} \qquad (7.4)$$

which is the expression required.*

Next, it is evident that r is an even periodic function of ψ, with period 2π, so that r can be developed into a Fourier cosine-series of the form

$$r = \tfrac{1}{2}A_0 + A_1 \cos \psi + A_2 \cos 2\psi + A_3 \cos 3\psi + \ldots .$$

We leave it to the reader to show that

$$r = a(1 - e \cos \phi),$$

and to deduce that

$$\frac{r}{a} = 1 + \tfrac{1}{2}e^2 - 2e\left\{J_1'(e)\frac{\cos \psi}{1} + J_2'(2e)\frac{\cos 2\psi}{2} + \ldots\right\} \qquad (7.5)$$

Again, $\theta - \psi$ is evidently an odd function of ψ, with period 2π, so that there is an expansion of the form

$$\theta - \psi = C_1 \sin \psi + C_2 \sin 2\psi + C_3 \sin 3\psi + \ldots$$

For the determination of the coefficients in this case, see Watson, "Bessel Functions," p. 554.

Ex. Show that

$$\frac{1}{1 - e \cos \phi} = 1 + 2\{J_1(e) \cos \psi + J_2(2e) \cos 2\psi + J_3(3e) \cos 3\psi + \ldots\}$$

Deduce that

$$\frac{1}{\sqrt{(1 - e^2)}} = 1 + 2\{J_1^2(e) + J_2^2(2e) + J_3^2(3e) + \ldots\}.$$

§ 111. *Critical length of a vertical rod.*

When a thin uniform elastic rod has its lower end clamped vertically, the vertical position of equilibrium is stable if

* Series of the type $\sum_{s=0}^{\infty} A_s J_{n+s}\{(n + s)x\}$, where A_s is independent of x, are called *Kapteyn* series (Watson, Ch. XVII).

the length of the rod is less than a certain critical length. But for a rod of this critical length, the vertical position is one of neutral equilibrium only, so that, if the upper end is slightly displaced and held fast until the rod is at rest, it will remain in the displaced position when released.* This will appear in what follows.

§ 112. Let l be the length of the rod, a the radius of its cross-section, w the weight per unit length. Put $\mathrm{I}=\frac{1}{4}\pi a^4$, and let E be Young's modulus for the material of which the rod is made.

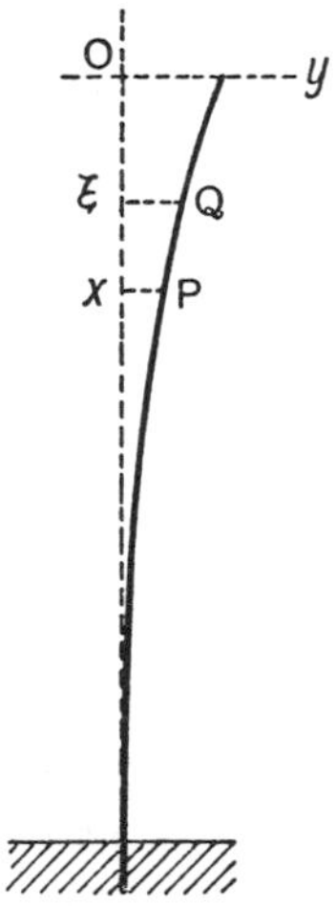

Suppose the rod to be in equilibrium in a position deviating slightly from the vertical (Fig. 19). Take the origin O at the upper end of the rod in the vertical position, the x-axis vertically downwards, and the y-axis in the plane of the rod. Let P be a point (x, y) on the rod, and Q a point (ξ, η) above P.

Fig. 19.

Consider the equilibrium of the part of the rod above P. The moment about P of the weight of an element $w d\xi$ at Q is $w\, d\xi(\eta - y)$, and by integration we obtain the moment about P of the weight of the part of the rod above P. Again, by the usual theory of elastic rods, the moment of the elastic forces about P is $\mathrm{EI}\, d^2y/dx^2$. Hence, since the part above P is in equilibrium,

$$\mathrm{EI}\frac{d^2y}{dx^2} = \int_0^x w(\eta - y)d\xi.$$

By differentiation with respect to x, we get

$$\mathrm{EI}\frac{d^3y}{dx^3} = [w(\eta - y)]_{\xi=z} - \int_0^x w\frac{dy}{dx}d\xi$$
$$= 0 - w\frac{dy}{dx}x,$$

* Greenhill: *Proc. Camb. Phil. Soc.*, IV, 1881.

that is,
$$\mathrm{EI}\frac{d^3y}{dx^3} + wx\frac{dy}{dx} = 0. \qquad (7.6)$$

Put $k^2 = w/\mathrm{EI}$; then
$$\frac{d^3y}{dx^3} + k^2x\frac{dy}{dx} = 0. \qquad (7.7)$$

Comparing this equation with (iv), § 105, we deduce that
$$\frac{dy}{dx} = \sqrt{x}\left\{\mathrm{AJ}_{-\frac{1}{3}}\left(\frac{2k}{3}x^{\frac{3}{2}}\right) + \mathrm{BJ}_{\frac{1}{3}}\left(\frac{2k}{3}x^{\frac{3}{2}}\right)\right\}, \qquad (7.8)$$

which can also be written, by expanding the Bessel functions,
$$\frac{dy}{dx} = a\left(1 - \frac{k^2x^3}{2\,.\,3} + \ldots\right) + bx\left(1 - \frac{k^2x^3}{3\,.\,4} + \ldots\right), \qquad (7.9)$$

where A, B or a, b are arbitrary constants.

Two conditions that must be satisfied by the particular solution required are :

(i) $d^2y/dx^2 = 0$ at $x = 0$, since there is no bending moment at the upper end ;

(ii) $dy/dx = 0$ at $x = l$.

Condition (i) gives $b = 0$. Condition (ii) can only be satisfied by $a = 0$, unless l satisfies the equation
$$0 = 1 - \frac{k^2l^3}{2\,.\,3} + \frac{k^4l^6}{2\,.\,3\,.\,5\,.\,6} - \frac{k^6l^9}{2\,.\,3\,.\,5\,.\,6\,.\,8\,.\,9} + \ldots, \qquad (7.10)$$

that is, the equation
$$\mathrm{J}_{-\frac{1}{3}}\left(\frac{2kl^{\frac{3}{2}}}{3}\right) = 0. \qquad (7.11)$$

Now * the least root of the equation $\mathrm{J}_{-\frac{1}{3}}(x) = 0$ is $x = 1{\cdot}8663$. Hence the rod cannot bend from the vertical until the length l is given by
$$\frac{2kl^{\frac{3}{2}}}{3} = 1{\cdot}8663,$$

and hence
$$l^3 = 7{\cdot}84\mathrm{EI}/w. \qquad (7.12)$$

* Gray and Mathews : "Bessel Functions," p. 317.

This gives the critical height of the rod. For example, for a steel rod of diameter 0·1 inch, density 0·28 lb. per cu. in., $E = 13{,}000$ tons per sq. in., we find $l = 80$ in., approximately.

§ 113. From a practical point of view, it is perhaps easier to solve equation (7.7) in series, and to solve (7.10) by trial. The solution is given here as an example on Bessel functions of fractional order.

Ex. In the problem of the small vibrations of a flexible string of length l with its ends fixed, the displacement y satisfies the differential equation

$$\frac{\partial^2 y}{\partial t^2} = \frac{T}{\rho}\frac{\partial^2 y}{\partial x^2}$$

where T is the tension and ρ the line density.

Show that the normal modes of vibration are given by

$$y = X \cos(\omega t - \epsilon),$$

where X is a function of x which satisfies the equation

$$\frac{d^2X}{dx^2} + \frac{\rho\omega^2 X}{T} = 0$$

and vanishes at $x = 0$ and $x = l$.

Hence, if $$\rho = \rho_0\left(1 + \frac{kx}{l}\right) = \rho_0\xi,$$

show that $$\frac{d^2X}{d\xi^2} + \kappa^2\xi X = 0, \quad \left(\kappa^2 = \frac{\rho_0\omega^2 l^2}{k^2 T}\right),$$

and deduce that the periods $2\pi/\omega$ of the normal modes are given by

$$\omega^2 = \frac{9\mu^2 k^2 T}{4\rho_0 l^2},$$

where μ is a root of the equation

$$J_{\frac{1}{3}}(\mu)J_{-\frac{1}{3}}(\lambda\mu) = J_{-\frac{1}{3}}(\mu)J_{\frac{1}{3}}(\lambda\mu),$$

and $\lambda = \sqrt{(1+k)^3}$.

§ 114. *Circular membrane with the circumference fixed. Normal modes of vibration.*

We return to the problem of the vibrations of a circular membrane with the circumference fixed, no longer assuming that the vibrations are independent of θ. With the same notation as in § 25, the differential equation to be satisfied is

$$\frac{\partial^2 z}{\partial t^2} = c^2\left(\frac{\partial^2 z}{\partial r^2} + \frac{1}{r}\frac{\partial z}{\partial r} + \frac{1}{r^2}\frac{\partial^2 z}{\partial \theta^2}\right). \quad . \quad . \quad . \quad (7.13)$$

To find the normal modes of vibration, we try a solution of the form

$$z = \mathrm{R}\Theta \cos(\omega t - \epsilon),$$

where R, Θ are functions of r, θ only. The result of the substitution can be written in the form

$$\left(\frac{d^2\mathrm{R}}{dr^2} + \frac{1}{r}\frac{d\mathrm{R}}{dr} + \frac{\omega^2}{c^2}\mathrm{R}\right)\frac{r^2}{\mathrm{R}} = -\frac{1}{\Theta}\frac{d^2\Theta}{d\theta^2}, \quad . \quad (7.14)$$

which is only possible if each side of this equation is equal to the same constant, since the variables r, θ are independent. Putting both sides equal to n^2, we deduce that Θ and R respectively satisfy the equations

$$\frac{d^2\Theta}{d\theta^2} = -n^2\Theta, \quad . \quad . \quad . \quad (7.15)$$

$$\frac{d^2\mathrm{R}}{dr^2} + \frac{1}{r}\frac{d\mathrm{R}}{dr} + \left(\frac{\omega^2}{c^2} - \frac{n^2}{r^2}\right)\mathrm{R} = 0. \quad . \quad (7.16)$$

From (7.15) we then have

$$\Theta = \mathrm{C}\sin(n\theta - \beta),$$

where C, β are arbitrary constants. Now, if the membrane is subject to no constraining force except that at the circumference, z must be a single-valued function of position and so must be of period 2π in θ. Hence n must be an integer, which, without loss of generality, we may take to be a positive integer, or zero.

The general solution of (7.16) can then be written

$$\mathrm{R} = \mathrm{AJ}_n\left(\frac{\omega r}{c}\right) + \mathrm{BY}_n\left(\frac{\omega r}{c}\right),$$

and as the particular solution required for the present problem is plainly one that remains finite when $r \to 0$, we must put $\mathrm{B} = 0$, since $\mathrm{Y}_n(x) \to \infty$ when $x \to 0$.

Hence, merging A and C into one constant, we have the solution

$$z = \mathrm{AJ}_n\left(\frac{\omega r}{c}\right) \sin (n\theta - \beta) \cos (\omega t - \epsilon), \quad . \quad (7.17)$$

where A, β, ϵ are arbitrary constants, and n is any positive integer, or zero.

Further, if $z = 0$ at $r = a$, for all values of θ and t, the equation

$$\mathrm{J}_n\left(\frac{\omega a}{c}\right) = 0 \quad . \quad . \quad . \quad (7.18)$$

must be satisfied by ω. Accordingly,

$$\frac{\omega a}{c} = \alpha_1, \alpha_2, \alpha_3, \ldots$$

where, since ω is positive, $\alpha_1, \alpha_2, \alpha_3, \ldots$ denote the positive roots of the equation $\mathrm{J}_n(x) = 0$. Consequently, the normal modes of vibration are given by

$$z = \mathrm{AJ}_n\left(\frac{r\alpha}{a}\right) . \sin (n\theta - \beta) . \cos \left(\frac{c\alpha t}{a} - \epsilon\right), \quad (7.19)$$

where α is any positive root of $\mathrm{J}_n(x) = 0$.

§ 115. The normal modes when $n = 0$ have been discussed in § 25. For the discussion of any other single mode, the constants β, ϵ are of no importance, as they depend only upon the initial line from which θ is measured, and the instant from which t is measured. Consequently, any normal mode for which $n \neq 0$ may be written in the form

$$z = \mathrm{AJ}_n\left(\frac{r\alpha}{a}\right) . \sin n\theta . \cos \frac{c\alpha t}{a}. \quad . \quad . \quad (7.20)$$

This represents a doubly infinite system of normal modes, for there is an infinite number of values of n, and an infinite number of values of α for each value of n.

When $n = 1$ we have

$$z = \mathrm{AJ}_1\left(\frac{r\alpha_s}{a}\right) . \sin \theta . \cos \frac{ct\alpha_s}{a},$$

where $J_1(\alpha_s) = 0$, $(s = 1, 2, 3, \ldots)$. In each mode of this set, the radii $\theta = 0$, $\theta = \pi$ together form a nodal diameter. For $s = 2$, there is one nodal circle, for $s = 3$ two nodal circles, and so on (Figs. 20.1, 20.2, 20.3).

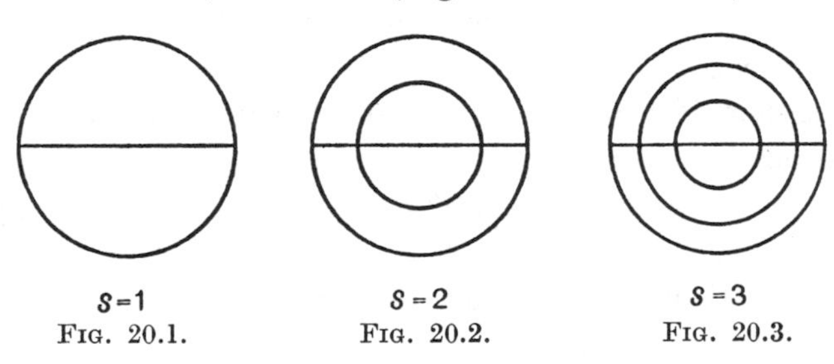

FIG. 20.1. FIG. 20.2. FIG. 20.3.

When $n = 2$ we have

$$z = AJ_2\left(\frac{r\alpha_s}{a}\right) . \sin 2\theta . \cos \frac{ct\alpha_s}{a},$$

where $J_2(\alpha_s) = 0$, $(s = 1, 2, 3, \ldots)$. In each of these modes the radii $\theta = 0$, $\theta = \pi$ together form one nodal

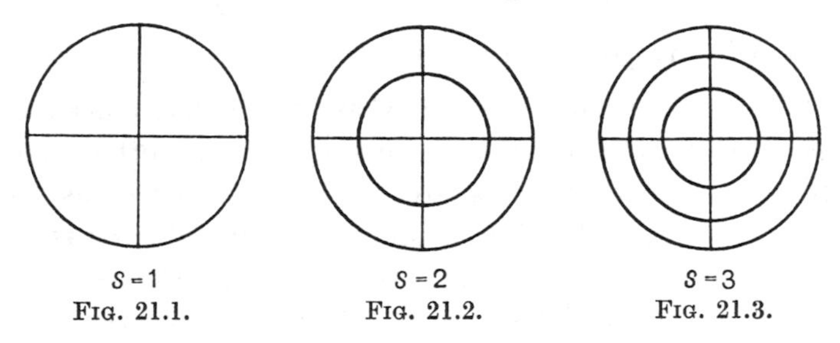

FIG. 21.1. FIG. 21.2. FIG. 21.3.

diameter, and the radii $\theta = \frac{1}{2}\pi$, $\theta = \frac{3}{2}\pi$ form another (Figs. 21.1, 21.2, 21.3).

§ 116. *General initial conditions.*

The most general solution that can be obtained by adding together arbitrary multiples of the normal modes may be written

$$z = \sum_{n=0}^{\infty} \sum_{s=1}^{\infty} AJ_n\left(\frac{r\alpha_s}{a}\right) . \sin (n\theta - \beta) . \cos \left(\frac{ct\alpha_s}{a} - \epsilon\right), \quad (7.21)$$

where α_s denotes the sth positive root of $J_n(x) = 0$. The suffixes n, s might have been appended to the arbitrary constants A, β, ϵ, but have been omitted to lighten the appearance of the formula.

These arbitrary constants can be chosen to satisfy general initial conditions such as

$$z = \phi(r, \theta), \quad t = 0, \qquad (7.22)$$

$$\partial z/\partial t = \psi(r, \theta), \quad t = 0. \qquad (7.23)$$

§ 117. In particular, if the membrane is started from rest, with an initial displacement given by $z = \phi(r, \theta)$, we have to satisfy the conditions

$$z = \phi(r, \theta), \quad t = 0, \qquad \text{(i)}$$

$$\partial z/\partial t = 0, \qquad t = 0. \qquad \text{(ii)}$$

Condition (ii) is satisfied by putting $\epsilon = 0$ in every term, and we may then write the solution, with a slight change of notation, in the form

$$z = \sum_{n=0}^{\infty} \sum_{s=1}^{\infty} J_n\left(\frac{r\alpha_s}{a}\right)(A_{n, s} \cos n\theta + B_{n, s} \sin n\theta) \cos \frac{ct\alpha_s}{a}.$$

Putting $t = 0$, we see that, in order to satisfy (i), the coefficients A, B must be determined from the expansion

$$\phi(r, \theta) = \sum_{n=0}^{\infty} \sum_{s=1}^{\infty} J_n\left(\frac{r\alpha_s}{a}\right)(A_{n, s} \cos n\theta + B_{n, s} \sin n\theta).$$

To find $A_{n, s}$, multiply both sides by

$$J_n\left(\frac{r\alpha_s}{a}\right) . \cos n\theta . r\, dr\, d\theta,$$

and integrate over the membrane. Then every term on the right vanishes except the one that involves $A_{n, s}$ and we get

$$\int_0^a \int_0^{2\pi} r\phi(r, \theta)\, J_n\left(\frac{r\alpha_s}{a}\right) \cos n\theta\, dr\, d\theta$$

$$= A_{n, s} \int_0^a \int_0^{2\pi} r J_n^2\left(\frac{r\alpha_s}{a}\right) \cos^2 n\theta\, dr\, d\theta$$

$$= \pi A_{n,\,s}\int_0^a r J_n^{\,2}\left(\frac{r\alpha_s}{a}\right) dr$$

$$= \pi A_{n,\,s}\int_0^1 ax \,.\, J_n^{\,2}(x\alpha_s) \,.\, a\,dx$$

$$= \tfrac{1}{2}\pi a^2 A_{n,\,s}\, J^2_{n+1}(\alpha_s),$$

as in (6.63). This determines $A_{n,\,s}$. To find $B_{n,\,s}$ we first multiply the expansion throughout by

$$J_n\left(\frac{r\alpha_s}{a}\right) . \sin n\theta \,.\, r\,dr\,d\theta,$$

and integrate over the membrane as before.

Examples XVIII

1. Show how to find the periods of the normal modes of vibration of a membrane in the form of a sector of a circle of radius a and angle π/m, the condition $z = 0$ being satisfied at every point of the boundary.

Also show how to find the solution which satisfies the arbitrary initial conditions

$$z = \phi(r,\,\theta), \quad (0 < r < a, \quad 0 < \theta < \pi/m),$$
$$\partial z/\partial t = \psi(r,\,\theta), \quad (0 < r < a, \quad 0 < \theta < \pi/m).$$

2. Investigate the normal modes of vibration of a complete circular membrane of radius a, with the condition $z = 0$ satisfied all round the circumference and also along the radii $\theta = 0$, $\theta = 2\pi$ (a sector of angle 2π). Show that in the simplest case, in which there are no nodal radii,

$$z = \frac{A}{\sqrt{r}} \sin\frac{s\pi r}{a} \sin\frac{\theta}{2} \cos\frac{s\pi ct}{a},$$

where s is an integer.

[Note that in this case the nodal circles divide any radius into equal parts.]

3. Show that the normal modes of vibration of a sector of a circle of radius a and angle 120°, with the condition $z = 0$ satisfied at every point of the boundary, in the simplest case, in which there are no nodal radii, are given by

$$z = \frac{A}{\sqrt{r}}\left(\frac{a}{r\alpha}\sin\frac{r\alpha}{a} - \cos\frac{r\alpha}{a}\right)\sin\frac{3\theta}{2}\cos\frac{ct\alpha}{a},$$

where α denotes a typical positive root of the equation $x = \tan x$.

4. Discuss the normal modes of vibration of a membrane in the form of a circular annulus bounded by concentric circles of radii a, b, with the condition $z = 0$ satisfied at every point of both circles.

5. In the problem of the circular membrane of radius a (§ 26), vibrating with circular symmetry and satisfying the condition $z = 0$ at $r = a$, if the initial conditions are

$$z = C\left(1 - \frac{r^2}{a^2}\right)^p, \quad \frac{\partial z}{\partial t} = 0,$$

where $p > 0$, show that

$$z = C \,.\, 2^{p+1}\Gamma(p + 1)\sum_{\alpha} \frac{J_{p+1}(\alpha)}{\alpha^{p+1}J_1^{\,2}(\alpha)} J_0\left(\frac{\alpha r}{a}\right) \cos\frac{\alpha ct}{a},$$

where $J_0(\alpha) = 0$.

6. In the problem of the uniform flexible hanging chain of length l, making small oscillations in a vertical plane (§ 27), if the initial conditions are

$$y = a\left(1 - \frac{x}{l}\right) + b\left(1 - \frac{x}{l}\right)^2, \quad \frac{\partial y}{\partial t} = 0,$$

show that

$$y = 8\sum_{\alpha} \frac{\alpha^2 a - 2(\alpha^2 - 8)b}{\alpha^5 J_1(\alpha)} J_0\left(\alpha\sqrt{\frac{x}{l}}\right) \cos\left(\frac{\alpha t}{2}\sqrt{\frac{g}{l}}\right),$$

where $J_0(\alpha) = 0$.

7. Show that a solution of the equation

$$\frac{\partial^2 v}{\partial r^2} + \frac{1}{r}\frac{\partial v}{\partial r} - \frac{v}{r^2} + \frac{\partial^2 v}{\partial z^2} = 0,$$

which satisfies the boundary conditions,

$$v = 0 \text{ when } z = 0 \text{ and when } r = a,$$

$$v = \frac{Vr(a^2 - r^2)}{a^3} \text{ when } z = l,$$

is

$$v = 16V\sum_{\alpha} \frac{J_1\left(\frac{\alpha r}{a}\right) \sinh\frac{\alpha z}{a}}{\alpha^3 J_2(\alpha) \sinh\frac{\alpha l}{a}},$$

where $J_1(\alpha) = 0$.

8. Laplace's Equation (§ 39) in cylindrical co-ordinates, r, θ, z is

$$\frac{\partial^2 u}{\partial r^2} + \frac{1}{r}\frac{\partial u}{\partial r} + \frac{1}{r^2}\frac{\partial^2 u}{\partial \theta^2} + \frac{\partial^2 u}{\partial z^2} = 0.$$

Show that the solutions of the form $u = f(z)g(r)h(\theta)$, which are everywhere finite and single-valued functions of position, are given by

$$u = (\mathrm{A}e^{-\mu z} + \mathrm{B}e^{\mu z})\mathrm{J}_n(\mu r)(\mathrm{C} \sin n\theta + \mathrm{D} \cos n\theta),$$
$$u = (\mathrm{A} \sin \mu z + \mathrm{B} \cos \mu z)\mathrm{I}_n(\mu r)(\mathrm{C} \sin n\theta + \mathrm{D} \cos n\theta),$$
$$u = (\mathrm{A}z + \mathrm{B})r^n(\mathrm{C} \sin n\theta + \mathrm{D} \cos n\theta),$$

where n is a positive integer or zero.

9. If U, V are two solutions of Laplace's equation, both continuous and one-valued throughout a region bounded by a simple closed surface, it is known from Green's theorem that

$$\iint \left(\mathrm{U}\frac{\partial \mathrm{V}}{\partial \nu} - \mathrm{V}\frac{\partial \mathrm{U}}{\partial \nu}\right) d\mathrm{S} = 0,$$

where $\partial/\partial\nu$ denotes differentiation in the direction of the outward (or inward) normal to the surface, and the integral is taken over the surface.

Prove (6.54) when n is a positive integer, by putting

$$\mathrm{U} = e^{-\alpha z}\, \mathrm{J}_n(\alpha r) \cos n\theta, \quad \mathrm{V} = e^{-\beta z}\, \mathrm{J}_n(\beta r) \cos n\theta,$$

integrating over the surface of the cylinder bounded by $r = 1$, $z = 0$, $z = l$, and making $l \to +\infty$.

10. Solve the problem of § 40 when the boundary conditions are $u = 0$, $(z = 0,\ 0 < r < a)$; $u = 0$, $(r = a,\ 0 < z < l)$; $u = \phi(r, \theta)$, $(z = l,\ 0 < r < a)$.

11. Solve the problem of § 40 when the boundary conditions are $u = 0$, $(z = 0,\ 0 < r < a)$; $u = 0$, $(z = l,\ 0 < r < a)$; $u = f(z, \theta)$, $(r = a,\ 0 < z < l)$.

[This requires a double Fourier series.]

12. Solve the problem of § 40 when the boundary conditions are $u = \psi(r, \theta)$, $(z = 0,\ 0 < r < a)$; $u = \phi(r, \theta)$, $(z = l,\ 0 < r < a)$; $u = f(z, \theta)$, $(r = a,\ 0 < z < l)$.

INDEX.

A CATALOG OF SELECTED

DOVER BOOKS

IN SCIENCE AND MATHEMATICS

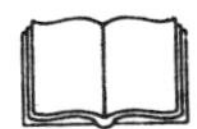

DE RE METALLICA, Georgius Agricola. The famous Hoover translation of greatest treatise on technological chemistry, engineering, geology, mining of early modern times (1556). All 289 original woodcuts. 638pp. 6¾ × 11.
60006-8 Pa. $17.95

SOME THEORY OF SAMPLING, William Edwards Deming. Analysis of the problems, theory and design of sampling techniques for social scientists, industrial managers and others who find statistics increasingly important in their work. 61 tables. 90 figures. xvii + 602pp. 5⅜ × 8½. 64684-X Pa. $15.95

THE VARIOUS AND INGENIOUS MACHINES OF AGOSTINO RAMELLI: A Classic Sixteenth-Century Illustrated Treatise on Technology, Agostino Ramelli. One of the most widely known and copied works on machinery in the 16th century. 194 detailed plates of water pumps, grain mills, cranes, more. 608pp. 9 × 12. (EBE)
25497-6 Clothbd. $34.95

LINEAR PROGRAMMING AND ECONOMIC ANALYSIS, Robert Dorfman, Paul A. Samuelson and Robert M. Solow. First comprehensive treatment of linear programming in standard economic analysis. Game theory, modern welfare economics, Leontief input-output, more. 525pp. 5⅜ × 8½. 65491-5 Pa. $13.95

ELEMENTARY DECISION THEORY, Herman Chernoff and Lincoln E. Moses. Clear introduction to statistics and statistical theory covers data processing, probability and random variables, testing hypotheses, much more. Exercises. 364pp. 5⅜ × 8½. 65218-1 Pa. $9.95

THE COMPLEAT STRATEGYST: Being a Primer on the Theory of Games of Strategy, J.D. Williams. Highly entertaining classic describes, with many illustrated examples, how to select best strategies in conflict situations. Prefaces. Appendices. 268pp. 5⅜ × 8½. 25101-2 Pa. $6.95

MATHEMATICAL METHODS OF OPERATIONS RESEARCH, Thomas L. Saaty. Classic graduate-level text covers historical background, classical methods of forming models, optimization, game theory, probability, queueing theory, much more. Exercises. Bibliography. 448pp. 5⅜ × 8¼. 65703-5 Pa. $12.95

CONSTRUCTIONS AND COMBINATORIAL PROBLEMS IN DESIGN OF EXPERIMENTS, Damaraju Raghavarao. In-depth reference work examines orthogonal Latin squares, incomplete block designs, tactical configuration, partial geometry, much more. Abundant explanations, examples. 416pp. 5⅜ × 8¼.
65685-3 Pa. $10.95

THE ABSOLUTE DIFFERENTIAL CALCULUS (CALCULUS OF TENSORS), Tullio Levi-Civita. Great 20th-century mathematician's classic work on material necessary for mathematical grasp of theory of relativity. 452pp. 5⅜ × 8½.
63401-9 Pa. $9.95

VECTOR AND TENSOR ANALYSIS WITH APPLICATIONS, A.I. Borisenko and I.E. Tarapov. Concise introduction. Worked-out problems, solutions, exercises. 257pp. 5⅜ × 8¼. 63833-2 Pa. $6.95

CATALOG OF DOVER BOOKS

CHALLENGING MATHEMATICAL PROBLEMS WITH ELEMENTARY SOLUTIONS, A.M. Yaglom and I.M. Yaglom. Over 170 challenging problems on probability theory, combinatorial analysis, points and lines, topology, convex polygons, many other topics. Solutions. Total of 445pp. 5⅜ × 8½. Two-vol. set.
Vol. I 65536-9 Pa. $6.95
Vol. II 65537-7 Pa. $6.95

FIFTY CHALLENGING PROBLEMS IN PROBABILITY WITH SOLUTIONS, Frederick Mosteller. Remarkable puzzlers, graded in difficulty, illustrate elementary and advanced aspects of probability. Detailed solutions. 88pp. 5⅜ × 8½.
65355-2 Pa. $3.95

EXPERIMENTS IN TOPOLOGY, Stephen Barr. Classic, lively explanation of one of the byways of mathematics. Klein bottles, Moebius strips, projective planes, map coloring, problem of the Koenigsberg bridges, much more, described with clarity and wit. 43 figures. 210pp. 5⅜ × 8½. 25933-1 Pa. $5.95

RELATIVITY IN ILLUSTRATIONS, Jacob T. Schwartz. Clear nontechnical treatment makes relativity more accessible than ever before. Over 60 drawings illustrate concepts more clearly than text alone. Only high school geometry needed. Bibliography. 128pp. 6⅛ × 9¼. 25965-X Pa. $5.95

AN INTRODUCTION TO ORDINARY DIFFERENTIAL EQUATIONS, Earl A. Coddington. A thorough and systematic first course in elementary differential equations for undergraduates in mathematics and science, with many exercises and problems (with answers). Index. 304pp. 5⅜ × 8½. 65942-9 Pa. $7.95

FOURIER SERIES AND ORTHOGONAL FUNCTIONS, Harry F. Davis. An incisive text combining theory and practical example to introduce Fourier series, orthogonal functions and applications of the Fourier method to boundary-value problems. 570 exercises. Answers and notes. 416pp. 5⅜ × 8½. 65973-9 Pa. $9.95

THE THEORY OF BRANCHING PROCESSES, Theodore E. Harris. First systematic, comprehensive treatment of branching (i.e. multiplicative) processes and their applications. Galton-Watson model, Markov branching processes, electron-photon cascade, many other topics. Rigorous proofs. Bibliography. 240pp. 5⅜ × 8½. 65952-6 Pa. $6.95

AN INTRODUCTION TO ALGEBRAIC STRUCTURES, Joseph Landin. Superb self-contained text covers "abstract algebra": sets and numbers, theory of groups, theory of rings, much more. Numerous well-chosen examples, exercises. 247pp. 5⅜ × 8½. 65940-2 Pa. $6.95

Prices subject to change without notice.
Available at your book dealer or write for free Mathematics and Science Catalog to Dept. GI, Dover Publications, Inc., 31 East 2nd St., Mineola, N.Y. 11501. Dover publishes more than 175 books each year on science, elementary and advanced mathematics, biology, music, art, literature, history, social sciences and other areas.